The Promise and Peril of CRISPR

The Promise and Peril of CRISPR

Edited by Neal Baer

 JOHNS HOPKINS UNIVERSITY PRESS BALTIMORE

Essays by Neal Baer; Rachel M. West and Gigi Kwik Gronvall; Marcy Darnovsky
and Katie Hasson; Kevin Doxzen and Jodi Halpern; Carol Padden and Jacqueline
Humphries; Ethan Weiss; Rosemarie Garland-Thomson; Peter F. R. Mills; R. Alta
Charo; and J. Benjamin Hurlbut appeared in a previous form in the Special Issue on
CRISPR in *Perspectives in Biology and Medicine* in 2020.

Johns Hopkins University Press
2715 North Charles Street
Baltimore, Maryland 21218
www.press.jhu.edu

Library of Congress Cataloging-in-Publication Data

Names: Baer, Neal, editor.
Title: The promise and peril of CRISPR / edited by Neal Baer.
Description: Baltimore : Johns Hopkins University Press, 2024. | Includes
    bibliographical references and index.
Identifiers: LCCN 2023049801 | ISBN 9781421449302 (paperback ; alk. paper) |
    ISBN 9781421449319 (ebook)
Subjects: MESH: Clustered Regularly Interspaced Short Palindromic Repeats |
    Gene Editing—ethics | Genetic Diseases, Inborn—therapy | Bioethical Issues
Classification: LCC QH442 | NLM QU 550.5.G47 | DDC 576.5—dc23/eng/20240108
LC record available at https://lccn.loc.gov/2023049801

A catalog record for this book is available from the British Library.

*Special discounts are available for bulk purchases of this book. For more information, please
contact Special Sales at specialsales@jh.edu.*

For Caleb

Contents

Introduction: Code Dread?   1
Neal Baer

Part One: Overview—The Era of CRISPR

1   CRISPR: Challenges Posed by a Dual-Use Technology   25
Rachel M. West and Gigi Kwik Gronvall

Part Two: Ethical Questions Raised by CRISPR Technology

2   Untangling CRISPR's Twisted Tales   45
Marcy Darnovsky and Katie Hasson

3   Heritable Genome Editing and International Human Rights   71
Kevin Doxzen and Jodi Halpern

4   Democratizing CRISPR: Opening the Door or Pandora's Box?   83
Ellen D. Jorgensen

5   Welcome to the CRISPR Zoo   92
Marcus Schultz-Bergin

Part Three: Personal Perspectives

6   Who Goes First?   103
Carol Padden and Jacqueline Humphries

7   Billie Idol   115
Ethan Weiss

8   Curing Cystic Fibrosis?   124
Sandra Sufian

Part Four: Diverse Voices

9   CRISPR and Gene Editing: Why Indigenous Peoples
and Why Now?   133
Krystal Tsosie

10   Do Trans/Humanists Dream of Electric Tits?
CRISPR and Transgender Bioethics   139
Florence Ashley

Part Five: The Dilemma of Controlling the Future

11   Velvet Eugenics: In the Best Interests of Our
Future Children?   153
Rosemarie Garland-Thomson

12   A Therapeutic Fallacy   172
Peter F. R. Mills

13   Genome Editing, in Time   185
Robert Sparrow

Part Six: Oversight and Monitoring

14   Regulating CRISPR: A Quest to Foster Safe, Ethical,
and Equitable Innovation   195
Andrew C. Heinrich

15   Should We Fear Heritable Genome Editing?   207
R. Alta Charo

16   Advancing Progressively Backwards: Guiding and Governing
Heritable Genome Editing   219
J. Benjamin Hurlbut

List of Contributors   245
Index   249

The Promise and Peril of CRISPR

Introduction

Code Dread?

Neal Baer

CRISPR keeps me up at night. I marvel at its potential to cure insidious genetic diseases and scourges like malaria. I shudder at the ways it might be misused to create biological weapons. What frightens me most, though, is what I cannot predict: How will we use CRISPR? Who will have the authority to monitor its use? How will it change evolution? How will it redefine the very nature of our existence?

CRISPR (clustered regularly interspaced short palindromic repeats) is an ingenious cut-and-paste system that homes in on a particular DNA gene sequence and snips it out using *Cas9* enzymes. That sequence is then replaced with a new one that rewrites or repairs it. Two different cells—*somatic*, from the body, and *germline*, from gametes—can be manipulated using CRISPR. Only germline genome cells can be inherited by future offspring.

As I write this introduction in early 2024, breakthroughs are rapidly occurring in CRISPR design and application, including research using CRISPR to target RNA with *Cas13* enzymes (known as *CrRNA*). Though we do not examine CrRNA in this volume, focusing instead on *heritable genome editing* (HGE), this new variety of CRISPR could be used to "control the expression of human genes in different ways—such as flicking a light switch to shut them off completely or by using a dimmer knob to partially turn down their activity" (New York University 2023, 2). A gene expression "modulator" could be used to treat diseases like Charcot-Marie-Tooth, where an overabundance of a protein is produced that causes peripheral muscular paralysis, or in treating RNA viral infections.

Using CRISPR to treat human disease began in 2019, when scientists in the United States used somatic editing to treat a patient with sickle cell disease. In Europe, scientists treated one with beta thalassemia. Both patients, suffering from genetic mutations of their hemoglobin genes, underwent bone marrow ablation. Then CRISPR was used to modify their blood-producing stem cells to make healthy fetal hemoglobin. The CRISPR cells were engrafted in each patient without the usual worry of donor rejection. Both patients are reported to be doing well, making healthy fetal hemo-globin. These CRISPR-based treatments have kept them from needing transfusions or hospitalization for severe anemia, vaso-occlusive crises, and strokes. Using CRISPR to treat these genetic mutations that cause inor-dinate suffering and a shortened life span is nothing short of miraculous, though who will pay for this expensive treatment will come to the forefront now that the FDA has approved a CRISPR treatment for sickle cell disease.

Why worry, then, about using CRISPR if it can be made safe in the clini-cal setting? Repairing the somatic genetic mutations in an individual means that scientists can do the same to the *germline genome*, and this could change human evolution by making an edit in an embryo that will be inherited by descendants. "That raises the possibility, more realistically than ever before," notes Michael Specter (2015), a well-known science journalist, "that scientists will be able to rewrite the fundamental code of life, with consequences for future generations that we may never be able to anticipate" (54).

The impact of CRISPR goes beyond the human species. Any gene can po-tentially be edited, whether in a person, animal, or plant—if it has a ge-nome. Imagine editing crops to be drought tolerant or animals, like pigs, to make organs that can be transplanted into humans (known as *xenotransplan-tation*). These are not ideas in the realm of science fiction; CRISPR is used today on plants and animals to accomplish these feats and much more. Re-cently, researchers edited genes in chicken cells by using CRISPR to create "chickens that can resist real-life doses of avian flu viruses." However, "the edited birds still became infected when exposed to larger amounts of the flu virus. And the strategy raises a safety concern: chickens edited this way could, in theory, drive the evolution of flu variants better at infecting people" (Co-hen 2023, para. 1, 2).

Even more astounding is the potential to eliminate malaria with a CRISPR-based gene-drive system that preferentially propagates a genetic trait through an organism and its offspring. Gene drives could be used to make a female mosquito infertile, causing it to be incapable of laying eggs. Instead of a 50% chance that the disruptive gene is inherited, a CRISPR gene copies itself, making the inheritance 100%, leading to mosquitoes that could extinguish their own species. The debate now focuses on whether to release these gene drives into malaria-ridden communities and whether scientists can ever truly predict the unintended ecological outcome of making a species extinct.

Can we ever be certain of the environmental impact of such an undertaking, even if it means preventing a thousand children from dying of malaria each day? Can we prevent the malevolent use of CRISPR to conjure up bacteria or viruses that can wipe out millions or destroy food crops the world depends on? "If you can edit a creature to solve a problem," declares gene-drive discoverer Kevin Esvelt (2019), "you can edit a creature to create a problem" (5).

CRISPR has vaulted us into a binary world of hope and potential devastation. It can provide cures that will alleviate suffering along with terrifying bioweapons that could destroy us. It can lead to somatic gene editing treatments for diseases that assail humankind and germline gene-editing enhancements that could give the wealthy or empowered an even greater societal edge. Political theorist Michael Sandel (2004) warns that this could lead to two subspecies: "the enhanced and the merely natural" (para. 13). This dual-use research of concern "could be directly misapplied by others to pose a threat to public health and safety, agricultural crops and other plants, animals, the environment, or materiel," or national security (National Research Council 2004, 17).

CRISPR is not the first exterminating peril the world has faced. Scientists and inventors have given us nuclear weapons and fossil-fuel engines that have posed dire threats of nuclear winter and global warming. Since the atomic bomb was dropped on Hiroshima and Nagasaki, nation-states have operated under the paradigm of mutually assured destruction: if you bomb us with a nuclear weapon, we will retaliate. The threat of mutual annihilation has averted nuclear Armageddon. With CRISPR, this old paradigm has evaporated.

Today, a rogue scientist with access to a DNA synthesizer, DNA sequences—which can be easily purchased—and some basic scientific expertise can build genetic weapons of mass destruction. Scientists have already made horsepox, polio, and virulent avian flu (Yong 2018). CRISPR has also been used to edit embryos, which were not brought to term. Then, in 2018, twin girls were born with a CRISPR-engineered germline mutation.

Dr. He Jiankui, a Chinese geneticist, used CRISPR to manipulate the germline of embryos that became the twins, claiming to have made them resistant to HIV. An uproar ensued against He among scientists, journalists, and bioethicists. A few, however, like the Harvard geneticist George Church, supported him, claiming in an interview that He was subjected to "bullying" and that "as long as these are normal, healthy kids it's going to be fine for the field and the family" (Cohen 2018, para. 6). At the end of 2019, He and his two colleagues were sentenced to three years in prison and fined $430,000 for illegally practicing medicine by "knowingly violating [China's] regulations and ethical principles with their experiments. . . . The court indicated that three genetically edited babies have been born" (Kennedy 2019, para. 4).

Where will the CRISPR path lead us? That is a key question pursued in *The Promise and Peril of CRISPR*.

Before and After

Dr. He's manipulation of the twin embryos has changed how scientists and the public view CRISPR. Before his experiment, HGE was regarded as verboten by numerous academies and scientists, at least until strict safeguards could be put in place. These twins raise many questions about moving forward with germline editing. What was conceivable has been made real.

Meiosis is nature's way of sorting and re-sorting the genes of most living organisms, guaranteeing diversity within species and driving evolution to select attributes—in the form of mutations—that benefit the organism and favor survival. Those mutations also inflict some individuals with painful, life-limiting genetic syndromes. Some, like Tay-Sachs, are lethal early in life, and others, like Huntington's, are genetic time bombs. CRISPR can seek out and destroy genetic scourges, somatically and heritably.

We must address the ethical dilemmas CRISPR raises, particularly when we begin to genetically tinker with the germline of human beings and any other living plant or organism. Scientists, agronomists, and even dog and cat breeders have been manipulating the germline of species for thousands of years, but not until recently has it been possible to do this in a specific gene-targeted way. CRISPR has drawbacks because it can cause off-target mutations, but newer CRISPR-based tools, like prime editing—"a versatile and precise genome editing method that directly writes new genetic information into a specified DNA site"—offer the possibility of avoiding the errors that CRISPR can make (Anzalone et al. 2019, 149).

Considering these quickening advancements in CRISPR technology, we have collected 16 essays in this volume devoted to investigating the trenchant ethical issues that CRISPR poses. We have engaged renowned geneticists, bioethicists, and philosophers to write about what they think are the most pressing questions CRISPR raises today. It is our sincerest hope that you will not only come away with a deeper understanding of CRISPR but that you will raise your voice in how it should be used in the future.

CRISPR: A Rose Bearing Thorns

The essays in this collection probe the challenging questions posed by a gene-editing tool that can cure agonizing diseases but can also give an individual the power, as Esvelt (2019) warns, to "single handedly genetically engineer a whole species" (2). Scarcely anyone would stop the promising cures we have seen using somatic gene editing for sickle cell disease and other rare genetic blood, liver, and eye disorders—if the technology is safe, as has been the case in the few diseases CRISPR has been used to treat. HGE is another matter and is the focus of this volume.

To systematically cover the multitude of ethical questions posed by using CRISPR to edit germlines, we have divided the volume into six sections that raise recurrent themes throughout:

Overview: The Era of CRISPR
Is it possible to reconcile the dual uses of CRISPR technology?
How will our present-day decisions about using heritable genome editing
    affect the future when the future is not predictable?
What social, political, and economic forces drive the development of
    CRISPR technology?

Ethical Questions
What are the dominant narratives that permeate discussions about CRISPR?
How ought we proceed in using HGE and who should be included in the
    debate?

Personal Perspectives
Should HGE be used to edit disabilities and lethal genetic syndromes, and
    what constitutes a disability?
How might we understand a mother's reproductive freedom versus a
    fetus's future health?

Diverse Voices
Who will be empowered to decide what genetic features are deemed desir-
    able or undesirable, and will that further disenfranchise minorities?

Should CRISPR technology be used to enhance one's physical desires, say in the case of trans individuals?

Dilemma of Controlling the Future

Should HGE be used in the exceedingly rare case when no other technology exists to ensure that a child is biologically related to both parents? What is the future impact of CRISPR on human variability?

Controls, Oversight, and Monitoring

Who should have authority over decisions about whether, when, and how CRISPR should be used?
How might we oversee who has access to CRISPR to prevent its misuse?

Dual-Use Technology: The Good, the Bad, and the Ugly

Chapter 1, by Rachel West and Gigi Gronvall, sets the stage with an overview of the historical development of CRISPR technology, describing "the expansion and refinement of CRISPR as a genetic engineering tool." The authors then present a discussion of CRISPR's dual use, "that, in addition to its numerous benefits, it also has the potential to be misused and could lower technical barriers to biological weapons development."

Considering the challenges of CRISPR's dual use, West and Gronvall outline "potential biosecurity concerns; and [recommend] steps governments and scientists may take to reduce biosecurity risks." They note that we must come to terms with the possibility that CRISPR could be used for nefarious purposes, where a rogue scientist or highly trained amateur could "edit an existing pathogen to make it more damaging, edit a nonpathogenic organism to incorporate pathogen genes and traits, and even, theoretically synthesize a novel pathogen." In light of these risks, and that "there are no 'total' solutions that can reduce the risks of misuse to zero," the authors offer safeguards and "partial solutions" that governmental agencies, science academies, health governance organizations, and companies dealing with DNA sequences can put in place to mitigate catastrophic scenarios. A good place to start, as several of our authors point out, is in high school and college classrooms and in do-it-yourself (DIY) community labs, by embedding ethics discussions into the science curricula and instruction (discussed in this volume by Heinrich and Jorgensen). The goal is encouraging students to develop and nurture the habit of thinking about ethics along with biosafety (Karoff 2019).

The Stories We Tell

How we think about CRISPR is determined by many social, cultural, political, and economic factors. Scientists and physicians tell stories differently from individuals without the experience of researching genetic syndromes or treating patients suffering from them. Marcy Darnovsky and Katie Hasson encourage us to refocus our conversations about CRISPR to "explore the considerable misrepresentations and distortions that skew public understanding" and to "foster constructive conversations about [HGE] and make sure they are considered in consequential decision-making processes." They point out that one problem with the stories we tell is that they are often infused with emotion that "shapes our thinking about how gene-editing tools should and should not be used." Terms like *designer* and *CRISPR babies* can evoke images of monsters or enhanced superhumans that "distort understandings of how, by whom, and to what ends HGE might be used." We must also be wary of personal anecdotes. Though often powerful, they can be misleading. Darnovsky and Hasson relate the story of a mother speaking at the First International Summit on Human Gene Editing in 2015, whose child died in infancy. She pleaded for HGE to move forward to save the lives of babies like hers. In reality, her baby suffered from anencephaly, which had "no clear genetic basis," yet that moving story was quoted numerous times in the media.

With so many stories being told about CRISPR, particularly why we should and should not move forward with HGE, Darnovsky and Hasson maintain that it is essential to "develop robust forms of public engagement" that should include "women's health advocates, disabled people and communities, and those working to reduce economic and racial inequalities," along with "scholars from the humanities and social sciences." One might add theologians, philosophers, and anyone else who wishes to learn more and hear about the prospects of genetic engineering. In this volume, we also bring in the voices of two bioethics scholars: one, a member of an Indigenous community and the other, a transfeminist. As members of the human community, we all should have a say in how CRISPR should be used. Bringing all parties to the table requires what I call *innovative conversations*.

To date, the debate on CRISPR and germline genome editing has been limited to scientific societies and bodies, with little public discussion among political candidates or from the White House. A topic as critical to our future must be discussed by everyone, not just experts. "Genuine inclusivity in these deliberations," Darnovsky and Hasson write, "will require challenging the medical framing that favors professional and technical authority, constrains the questions asked, and delimits which risks and benefits are seen as relevant." We must design new formats to promote innovative conversations to accompany innovative—and potentially disruptive—genetic technologies.

Pathway as Metaphor

After the births of the CRISPR-edited twins, the Organizing Committee of the Second International Summit on Human Genome Editing concluded that "the scientific understanding and technical requirements for clinical practice remain too uncertain and the risks too great to permit clinical trials of germline editing at this time. Progress over the last three years and the discussions at the current summit, however, suggest that it is time to define a rigorous, responsible translational pathway toward germline trials" (National Academies of Sciences, Engineering, and Medicine [NASEM] 2018, para 6). One of this volume's contributors, the bioethicist Peter Mills, has previously written that the translational "pathway is a powerful organizing metaphor for technology. The pathway drives traffic to its destination. If you have not asked for a pathway, or if you don't want a pathway, being presented with fully drawn-up plans for a pathway may seem a little presumptuous" (quoted in this volume by Darnovsky and Hasson). In other words, a pathway presumes that there is a goal—and that goal is HGE.

In their essay, Kevin Doxzen and Jodi Halpern note, with trepidation, that opening the path for HGE clinical research might allow it to proceed under some conditions. Nevertheless, they write that "while some authoritative bodies are focusing on HGE's safety and efficacy, we suggest an approach that begins with identifying serious concerns about social exclusion and social justice. These concerns . . . cannot be resolved by aiming to maximize positive over negative health outcomes. Rather, concerns raised by HGE reflect people's individual rights, rights that exist independently of population-level health outcomes." In contrast, George Church recently stated: "I believe we have an ethical obligation to maximize benefits and minimize harm. If we have an opportunity to eliminate infectious and genetic disease, or some subset of them—even one of them—then we should pursue it" (quoted in Hall 2019, para. 16). Can future generations, who will inherit whatever germline edits we make today, no matter how well intended, consent to these changes? Is it nonsense to think about unborn people consenting?

Doxzen and Halpern review several potential outcomes using HGE. They reflect on public access to medically necessary treatments and whether enhancements for one's prospective children should ever be allowed. They raise deep concerns about eugenics that can be conflated with our quest to improve our health and that of our future children. As national and international scientific bodies meet to develop guidelines and regulations for using HGE, Doxzen and Halpern argue that a human rights-based approach, rather than a utilitarian one, ought "to prioritize human rights when deciding permissible uses of HGE . . . consider[ing] the possibility that genome-editing technology . . . can improve society's collective health but focus[ing] first on existing risks to individuals' core rights."

Revisiting the Island of Dr. Moreau

Apart from bioethical discussions about human germline editing and the potentially nefarious uses of CRISPR to make bioweapons are experimental uses across species. H. G. Wells was prescient when he wrote *The Island of Dr. Moreau* in 1896. Dr. Moreau experimented by combining humans with animals. The ethicist Jason Scott Robert (2018) has written that "scientists have tried to 'humanize' their experimental model organisms (mouse, rat, pig, cow, non-human primate). With the advent of CRISPR-*Cas9*, the prospects for humanization have increased considerably. . . . With part-human chimeras, one ambition is to transfer human pluripotent stem cells into non-human embryos to assess the potential for cellular functional integration" (para. 5).

Amateur Dr. Moreaus can now purchase kits online that use CRISPR technology to increase the size of green tree frogs or make brewer's yeast fluoresce. Should biohackers be given the tools to conduct CRISPR experiments in their homes or community labs? Should DNA sequences be readily available to them for purchase?

Ellen Jorgensen, a cofounder of Genspace, the first nonprofit community biotechnology lab and a leader in promoting the democratization of CRISPR, argues that the "societal implications of discoveries are seldom clear at the beginning, and the public is normally not involved in shepherding new technologies into areas that align with their needs." From her unique perspective as one of the first "CRISPR teachers," Jorgensen provides a history of the citizen science movement and discusses her quest to help raise public awareness about CRISPR. By providing the public with the first DIY Biolab to learn CRISPR technology, Jorgensen hopes that hands-on training will "embed the societal impact discussions" in the activities undertaken by citizen scientists. With democratization and easy access to the materials needed to use CRISPR come biosecurity threats. Whether engaging the public in DIY "biohacking" will lead to citizen scientists addressing and voicing their views about the societal implications of heritable genome editing remains to be seen.

Biohacking—or hybridization—has long been used in the livestock, poultry, and agricultural industries to improve productivity and decrease the time it takes to raise animals, plants, and grains for food. With CRISPR comes new questions such as those raised by the philosopher Marcus Schultz-Bergin: will CRISPR improve animal welfare by decreasing suffering and improving health, or will it lead to "headless chickens" or "football birds" that wholly lack consciousness, effectively becoming "growing meat sacks"? The goal of gene editing for humans, Schultz-Bergin argues, has been "welfare promoting." But in the case of animals, "the history of both selective breeding and genetic engineering illustrates that our goals are typically at odds with the animal's welfare." How we reconcile using CRISPR on animals for our benefit may make it easier to use it in ways that may not promote the welfare of all human beings. This leads us to the next section on the personal stories that raise profound questions about editing out mutations in the human genome.

When the Personal Becomes Genetic

The *New York Times* science writer, David Wallace-Wells (2023), recently wrote that "four hundred million people worldwide are afflicted by one or more diseases arising from single-gene mutations that would be theoretically simple for CRISPR to fix" (para. 10). Somatic gene editing is in its infancy but is being tested on a wide range of diseases, including "congenital blindness, heart disease, diabetes, cancer and HIV" (Wallace-Wells 2023, para. 10). What if we decided to use HGE to permanently eliminate single-gene mutations from the human genome instead of treating individuals suffering from diseases like sickle cell anemia?

One of the arguments against using CRISPR to rewrite the germline is that options already exist for determining whether an embryo might carry a parent's genetic mutation. Parents can undergo genetic testing and, through in vitro fertilization (IVF) and preimplantation genetic diagnosis (PGD), implant embryos that do not contain that mutation. Others note that IVF can pose risks to the mother or woman donating eggs, and some object to destroying embryos on religious grounds (though one might presume that they would also object to using CRISPR to change an embryo that nature or God made). George Church, in a recent interview, says that we "should be focusing on outcomes rather than methods" when it comes to treating genetic diseases like sickle cell anemia, cystic fibrosis, and beta thalassemia. He concedes that "most of these can be cost-effectively dealt with by genetic counseling" and IVF, but he offers several arguments in favor of using HGE: it is less expensive to eliminate a genetic disease through the germline than to keep treating it somatically over generations; it is the best system for reaching all cells in the body; and it goes through a single cell rather than millions that involve somatic-cell treatments (quoted in Davies 2020).

What about exceptions, or what Peter Mills identifies as "the most unusual case"? Let us imagine that two parents, both homozygous for deafness, desire to have a child who would not be born deaf. That scenario is not imaginary. A Russian scientist has identified five deaf couples who want to have hearing children, and he has petitioned Russian authorities to allow him to carry out a CRISPR germline treatment on their embryos. This recent push to allow for HGE brings us to a discussion of using CRISPR germline editing by three contributors with personal experience.

Carol Padden, who is deaf, and her daughter, Jacqueline Humphries, who is not, propose an expansive view of what it means to be human, "to understand how to promote and not diminish genetic and cultural diversity." How one defines a disease, or a medically unmet need, is influenced by the culture and those in power, including the medical-industrial complex that is driven to find cures. Padden and Humphries point out that many in the deaf community do not see themselves as disabled but as valued contributors to human diversity, with a highly developed language and style of communication. Padden herself does not want to be "cured" because she is not inflicted, though she and Humphries are concerned "that the debate around who should first benefit from new medical breakthroughs favors expediency and urgency of scientific discovery. Deaf individuals," they argue, "may find themselves first in line for germline editing because their autosomal nonsyndromic recessive condition offers ideal clinical opportunities compared to those with more serious or debilitating diseases."

Ethan Weiss, in his essay about his daughter, Ruthie, born with albinism, offers that Ruthie has brought to his family "a perspective that we never could have had before she was here." He details the journey he and his wife, Palmer, have made in raising a child with a genetic disorder that severely impairs her vision, makes her intolerant to sunlight, and causes other medical problems. He also embraces Ruthie for who she is and the joy she brings to her family, friends, and those who meet her. Weiss also says, "beyond a doubt, that had we been aware of Ruthie's condition before she was born, she would not be here today." That knowledge is predicated on what he knew then; not on what he knows today. He worries that new technologies like CRISPR will stop us from having children like Ruthie in the future as we become better at detecting mutations and eliminating them. He leaves us with a conundrum: "one cannot know what it is like to parent a child with a disability until one knows. Without knowing, it is practically impossible to make an informed decision about whether and how to intervene."

Sandra Sufian, who was born with cystic fibrosis (CF), writes incisively that "part of living with a genetic disease like CF is the inability to delink my disease from myself; CF is an essential part of my being, not an appendage that one can splice away." Sufian poses thoughtful questions about what it means to have a disability, noting that having CF has provided her with a strong and vibrant community "form[ed] around genetic identity." Scientists and the medical industrial complex must be extraordinarily cautious in deciding what genetic diseases might be eliminated from the human genome by using HGE, and she rightly insists that we must examine "the fears and joys of real people's lives with CF" and other genetic diseases before promulgating "cures" for the unborn.

Welcoming Voices

The variety of voices in this volume includes those that have traditionally not been heard. Here we add two bioethicists: Krystal Tsosie, an Indigenous person, and Florence Ashley, a transfeminist, to deepen our understanding of the potential impact of HGE on those who have been underrepresented and underappreciated in the CRISPR debate.

Tsosie questions whether CRISPR treatments will be available to Indigenous peoples, looking back on the history of using their genetic data: "A glance at the failure of genetic innovation to impart direct benefit to Indigenous peoples, whose genetic data and knowledges have been usurped for the furtherance of academia and industry, should give critical pause as to whether gene-editing tools will truly democratize genomics." Tsosie is also deeply concerned that germline editing will be used as a "solution for defining what constitutes a normal human being. These are value-laden judgments that raise the danger of eugenics."

The question of who will be empowered to make these decisions is a theme raised throughout this volume, particularly by those with disabilities, those who have been disenfranchised, and those who identify as gender minorities. Tsosie raises profound questions that must be addressed: "Will certain genes associated with Indigenous peoples be judged less desirable and be replaced or modified using CRISPR?" Will those with political and economic power decide who should be genetically modified or enhanced?

Florence Ashley raises similar concerns regarding underrepresented groups in debates about CRISPR, in this case using HGE to try to modify embryos that might carry genes associated with transness. Ashley is concerned that scientists will search for and "identify genetic or hormonal causes for being trans" and "reinvigorate conversion practitioners in their crusade against trans existence." Considering today's chafing political atmosphere across many states and legislation preventing trans youth—and in some cases adults—from receiving gender-affirming medical care, Ashley forthrightly states that she "cannot fathom tolerating practices that try to prevent people from being trans or seek to eliminate them from society. Is that not, after all, what lies at the heart of the immorality of eugenics? Not just harm—although harm indeed—but inequality and dehumanization. Harm, not just to the individual, but to the very moral fabric of society."

On the other hand, Ashley presents an intriguing idea for using somatic gene editing to assist in transition. "Not only could it facilitate hormone-related care by replacing pills and injections with endogenous production, but it offers the promise of customization. . . . Personalized hormone regimens are still largely unknown in trans health care, with microdosing estrogen and testosterone still in their infancy outside of do-it-yourself spheres. While dosage can alter the effects of hormones, it is not yet possible to pick-and-choose results and some would like to change their bodies in ways that transcend what hormones allow." Will somatic gene editing further blur the binary cis world to what it means to be male or female, allowing trans and nonbinary (and even cis) individuals to select the traditionally masculine and feminine attributes they desire?

The Future Is Here—Or Is It?

Rosemarie Garland-Thomson presents a series of formidable questions CRISPR raises when we manipulate the genes in somatic and germline cells. While CRISPR is often discussed in terms of relieving human suffering,

Garland-Thomson asks us to address what we mean by suffering. Where do we draw the line on remediating disability when there is no consensus on that term? "The business of modern medicine . . . is to sort human variations into the opposing categories of disease and nondisease," writes Garland-Thomson, "or what has been considered since the nineteenth century as 'normal.' Making up new disease categories, or what medical science calls discovering new diseases, is a market-driven growth industry." Garland-Thomson asks us to think hard about how we define and categorize disease and health, lest we slip into a mindset of "liberal eugenics," that "enforces health as an unassailable aim [that] takes precedence over ethical interests."

It is a slippery slope where we accept CRISPR being used to treat whatever one feels is not healthy. When money is to be made, one can imagine CRISPR being used for enhancement for short stature or not having the musculature of a top athlete. Garland-Thomson continues that it is difficult "reimagining then through the knowledge of now," particularly when we think about whether one would have terminated a pregnancy or used CRISPR on a fetus born with a genetic syndrome, considering knowledge one has long after that child has been born. Yet with CRISPR potentially available for clinical use, she cautions us that we need "to make decisions in the present that will yield intended but uncertain future outcomes."

Many scientists contend that the driving question CRISPR poses is whether it fulfills an unmet medical need and can be made clinically safe. Garland-Thomson, along with many of our other contributors, warns that we must look further to "what the existence of CRISPR technology suggests about the limits of being human." Garland-Thomson worries that HGE approaches a "new eugenics." Using germline editing to enhance or improve future persons, she says, may lead to "morally unacceptable consequences, ranging from producing medical harm to abrogating consent, intensifying genetic discrimination, increasing social inequality, promoting conditional parental acceptance, turning people into products, fostering a commercial medical industrial complex, and encouraging rogue scientific and medical practice."

Garland-Thomson questions the faith we traditionally place in scientific breakthroughs, noting the "collateral damage" from mechanical inventions ranging from nuclear energy to gasoline-powered engines, plastic, or opioid medication—"all aimed at making a better future for everybody." It is not until many years later that the dire consequences of burning fossil fuels or dumping plastics into the ocean are being felt across the globe. Some may argue that by placing a moratorium on HGE, we are hedging our bets against the possibility that something may go horribly wrong and are therefore missing opportunities to alleviate suffering. But as Garland-Thomson and other essayists argue, the pathway to HGE is not a simple do or do not; it raises complex ethical issues that do not have simple answers.

Peter Mills seeks to clarify lines between what constitutes an unmet medical need and an enhancement. He argues that in the case of a couple carrying a genetic mutation and desiring a child, there is no unmet medical need because no child exists—that medical need can only exist once a child is born. He goes on to write that the "clinical trial model places too much weight on the patients' consent and interests, and it does not integrate the difficult question of the future person's interests, independent from the prospective parents' interests in their child's existence." Moreover, as Garland-Thomson argues in the case of Down syndrome, for instance, knowing that an embryo carries the extra chromosome for Down syndrome does not tell you what the child's life will be because there is a wide range of presentations of that syndrome. Determining that an embryo has a genetic mutation is not tantamount to knowing how that individual will experience the disease or how that individual will affect the lives of others. "This testing of what medical science understands as the *health* of the embryo or fetus," Garland-Thomson writes, "can put the best interests of a pregnant woman in conflict with those of her future child."

Robert Sparrow is also deeply interested in how we think about the future use of HGE and how our conceptions of that future direct our decisions on whether to engage in germline editing. "We imagine new technologies [as though] they have just been developed but are also, implausibly, fully realized. We seldom imagine them as old news. Rather, we conceive of our technological future primarily as a present yet to arrive." Problematically, we often do not account for improvements in the technology over time, he argues, although technological improvements in medical treatments are ongoing. Moreover, the "generation making decisions about the shape of the technology [seldom has] to live with the consequences. . . . We must understand [new technologies] as projects that have a temporal dimension, such that the ethical issues they raise alter as they move from the future to the present and then into the past."

Babies are not cell phones that can be discarded when new models—or improved treatments—arrive. Our genomes are, in many ways, the luck of the genetic dice—we have no control over which genes we will inherit from our parents along with mutations and traits that may confer physical and intellectual advantages or disadvantages. HGE changes the game of life. It may be possible to move from treatment to enhancement, giving control to those in power to confer advantages to their offspring and to edit out traits from the disenfranchised, potentially disenfranchising them more. How we conceive of HGE now will have profound, unknowable effects on future generations.

Sparrow concludes his essay by challenging us to think about how HGE "will blur the distinction between 'the born' and 'the made.' In the future, our nature—or at least our genes—will reflect the choices of others. This, in turn, may affect our ability to understand ourselves as the sole authors of our lives."

Regulation, Fear, and Progress

The concluding section of *The Promise and Peril of CRISPR* seeks to understand how this Promethean technology can be managed in a way that it does not fall into the wrong hands or is hijacked by a "rogue scientist." CRISPR's dual-use technology, with its somatic gene-editing cures, its potential for nefarious bioweapons use, and the thorny questions raised by germline editing, pose some of the most important questions facing society today. How should we control access to and regulate CRISPR? Is HGE to be feared or embraced? How can we fully answer these questions today? Is that even possible?

Andrew Heinrich, a lawyer and founder of Project Rousseau, a youth empowerment organization, raises these fundamental questions about regulatory oversight: "Is there an existing agency or framework that suits the new technology? Is there a combination thereof that might collectively serve the purpose? Or is an entirely new apparatus required?" To date, little federal legislative action has been taken to regulate CRISPR technology, which, arguably, suits many scientists who wish to conduct their research unperturbed by oversight and "red tape"—as one geneticist told me, "We need to police our own." Still, "only two passed and ratified federal statutes empower agencies to take action directly related to CRISPR, and the resulting agency action has been noncommittal." Moreover, a "far more striking trend is that the little legislative and regulatory action that has been taken does not address any of the central ethical or safety issues that constitute a majority of scientific and public concern about CRISPR technology."

This lack of involvement by the federal government does necessarily indicate disinterest on its part, Heinrich argues, but may result from CRISPR being "under a decade old [and] there simply has not been enough time for the government to gain CRISPR experts." Nevertheless, CRISPR research and clinical trials march on. Heinrich outlines a framework for regulating CRISPR research and its applications, focusing on:

Safety, training, and licensure: "requiring training and licensure to use the technology, restricting the facilities in which the technology can be used, restricting use to certain contexts in which the need outweighs the risk, requiring labeling and other notifications of risks";

Ethical considerations: "research on human germline cells and to the ethical definition of informed consent when risks are still underresearched";

Research and innovation: "any regulatory regime must encourage research and innovation . . . that may improve quality of life, especially those [innovations] that might not be the most inherently profitable"; and

Justice and equity: "it is of the utmost importance that research agendas and funding streams prioritize the uses of CRISPR that are most helpful to the underserved, both in the United States and globally. . . . Otherwise, development will simply perpetuate inequality."

History of Fear

Alta Charo encourages readers to examine the history of reproductive technologies before responding negatively to CRISPR and HGE. She notes that "often lacking in this debate has been an effort to look back at debates surrounding earlier advances in reproductive technologies, most of which have been accompanied by fears of eugenics, the loss of human dignity, and the disruption of parent–child relationships. While these advances have each had pockets of abusive uses, they have been integrated into modern life without bringing about wholesale destruction of society."

For Charo, focusing on the facts of how IVF and PGD have been used will help to reassure us that fears about the worst-case scenario—human cloning—have not happened. She does caution, however, that germline editing "is simply the latest rehearsal of what are fundamentally the same concerns around intolerance for diversity or imperfection" and that we must insist on good governance with broad consensus both in the United States and around the world to ensure safety, access, and equity.

The Progress Chameleon

Scientists do not work in a vacuum. They also have aspirations, drives, ethical beliefs, and egos that shape the way they think about conducting research and its impact on society. J. Benjamin Hurlbut, in our final essay, examines the forces that influenced Dr. He Jiankui, cautioning that the way science is done can lead to an "ill-begotten experiment" like the one performed by He. Underlying the CRISPR-edited twin outcome is what Hurlbut cites as "a familiar sequence: scientific knowledge generates technological applications which, in turn, produce societal impacts and consequences." Science and innovation race along, and then ethical questions come afterward. He points out that the First International Summit on Human Gene Editing in 2015 required "broad societal consensus," but "after the He story broke, the Hong Kong summit organizing committee abandoned that commitment and issued a statement reasserting science's jurisdiction over the future, declaring that it is 'time to define a rigorous, responsible translational pathway toward [germline genome editing clinical] trials.'" Hurlbut is deeply concerned by this shift, which mirrors the paradigm of science racing forward with society lagging behind. After He conducted his experiment, that notion was reinforced: what is done is done; now it is time for society to deal with it.

Hurlbut met He at a scientific meeting in Berkeley, California, in January 2017. Hurlbut's fascinating account of the forces placed on He to be the first scientist to use CRISPR to edit a human germline is profound. Hurlbut contends that He internalized contemporary views in biotechnology that "led him to believe that his experiment would elevate his status in the international scientific community, advance his country in the race for scientific and technological dominance, and drive scientific progress forward against the headwinds of ethical conservatism and public fear. Thus, far from rejecting the norms of his professional community, the ideas that influenced He are at once powerful and ubiquitous in the world of science."

During that meeting in Berkeley, He read about Robert Edwards, who secretly produced the first "test-tube baby," which Edwards revealed in the popular press before winning the Nobel Prize years later. Hurlbut notes that despite prohibitions against germline genome editing, He moved forward, deeply influenced by Edwards's story and by the conversations among scientists at that meeting whom He deeply respected. Dr. He came to believe that "major breakthroughs are driven by one or a couple of risk-taking scientists. Heroic scientific achievements are often initially controversial; someone must break the glass."

Breaking the glass is a powerful metaphor that drives many scientists. Hurlbut, in a stunning revelation, contends that He consulted and confided in dozens of scientists, "including well-known and respected American scientists. Most expressed support. In most cases, these supporters knew about the actual project. A small number affirmed HGE and the idea that some maverick would have to be the first to cross the Rubicon, but expressed reservations about the genetic target He had selected.   Only four expressed unequivocal opposition. Indeed, He told [Hurlbut] that one of his interlocutors, an accomplished genome-editing scientist at a leading US university, told him that when He commercialized his baby-making enterprise (because that is what one is expected to do with one's inventions in the world of biotechnology), he wanted to be the company's fifth employee. He did not want to be the second, or even the third or fourth, because the technology would still be controversial at that stage. He wanted to wait for the company to be developed enough to be established and accepted. That is, he wanted to wait for society to catch up."

Considering Hurlbut's disclosures, it is more imperative than ever for the National Academies of Sciences and International Commissions to take an extremely hard look at the culture that motivated Dr. He.

Hubris

CRISPR makes us imagine the unimaginable, or, at the very least, what was once the provenance of *Jurassic Park* science fiction. How will it change human evolution? Will it further divide us into the enhanced and the merely normal? Will it privilege medicine to a degree we never dreamed possible, wiping away scourges and devastating genetic diseases? Will we be able to harness the power of CRISPR and still protect humanity from rogue scientists who could wreak havoc, the likes of which we have never seen?

Today we are facing many challenges that beg for our attention: How will we address the growing calamity of global warming? How will we address the growing concerns about AI technology disrupting human enterprise as we know it? How will quantum computing reshape computers as we know and use them? How should we think about IVG (in vitro gametogenesis), the latest reproductive technology that may soon be able to make human egg and sperm cells from pluripotent stem cells? How will this change family dynamics? How will society respond to gay or lesbian couples having biological children together?

CRISPR is not the only daunting issue we face, and the onslaught of so many complex challenges may either divert our attention or paralyze us. Therefore, we offer this collection of essays in the hope that the complexity of CRISPR ethics is made clearer so we may all enter the conversation about its use, particularly for heritable genome editing.

All the essays in this collection share a driving question: What does it mean to be human today and how should we think about our future generations? The decisions we make today about manipulating the human germline genome will have a profound—and unpredictable—effect on our descendants. We often talk about how technology has changed the world before and after the invention of the Internet and cell phone, but these inventions changed the ways we interact with each other. They cannot change our genetic makeup. CRISPR is a game changer.

We are in for a ride that is not fully conceivable. These essays are an attempt to frame the questions that will have a lasting impact on humanity before something dire occurs. Are we smart enough, empathetic enough, and ethical enough to deal with the complex questions posed by CRISPR? Here I turn to the humanities for an answer. The arts, literature, and music all come from minds, hands, and hearts that were not genetically altered. Human ingenuity and compassion have made the world livable and given us hope in times of despair. Evolution has given us an endless assortment of unique human beings and their gifts. We must use these gifts to tell moving stories that clarify the ethical issues that CRISPR raises through literature, music, poetry, dance, and all modes of storytelling that make us human. Now that we have the power to shape our own evolution, we must think long and hard about how we want to use it.

REFERENCES

Anzalone, Andrew V., Peyton B. Randolph, Jessie R. Davis, Alexander A. Sousa, Luke W. Koblan, Jonathan M. Levy, Peter J. Chen, et al. 2019. "Search-and-Replace Genome Editing without Double-Strand Breaks or Donor DNA." *Nature* 576: 149–57.

Cohen, Jon. 2018. "'I Feel an Obligation to Be Balanced': Noted Biologist Comes to Defense of Gene Editing Babies." *Science*, November 28, 2018. https://www.sciencemag.org /news/2018/11/i-feel-obligation-be-balanced-noted-biologist-comes-defense-gene -editing-babies.

Cohen, Jon. 2023. "In 'Proof of Concept,' CRISPR-Engineered Chickens Shrug Off Flu." *Science*, October 10, 2023. https://www.science.org/content/article/proof-concept-crispr -engineered-chickens-shrug-flu.

Davies, Kevin. 2020. "From CRISPR Multiplexing to Pleistocene Park." *GEN*, January 2, 2020. https://www.genengnews.com/insights/from-crispr-multiplexing-to -pleistocene-park-2/.

Esvelt, Kevin. 2019. "We Can Change the DNA of an Entire Species in the Wild." Interview by Caterina Fake. *Should This Exist Podcast.* May 3, 2019. https://shouldthisexist.com /wp-content/uploads/2020/10/ste-episode-transcript_-gene-drive.pdf.

Hall, Ann. 2019. "Unboxing CRISPR." *Colloquy*, September 5, 2019. https://gsas.harvard .edu/news/unboxing-crispr.

Karoff, Paul. 2019. "Embedding Ethics in Computer Science Curriculum: Harvard Initiative Seen as a National Model." *Harvard John A. Paulson School of Engineering and Applied Sciences*, January 29, 2019. https://seas.harvard.edu/news/2019/01 /embedding-ethics-computer-science-curriculum.

Kennedy, Merrit. 2019. "Chinese Researcher Who Created Gene-Edited Babies Sentenced to 3 Years in Prison." *NPR*, December 30, 2019. https://www.npr.org/2019/12/30/792340177 /chinese-researcher-who-created-gene-edited-babies-sentenced-to-3-years-in-prison.

National Academies of Sciences, Engineering, and Medicine (NASEM). 2018. "On Human Genome Editing II." Statement by the Organizing Committee of the Second International Summit on Human Genome Editing, November 28, 2018. https://www .nationalacademies.org/news/2018/11/statement-by-the-organizing-committee-of -the-second-international-summit-on-human-genome-editing.

National Research Council. 2004. "Biotechnology Research in an Age of Terrorism." https://nap.nationalacademies.org/catalog/10827/biotechnology-research-in-an-age -of-terrorism.

New York University. 2023. "AI Combined with CRISPR Precisely Controls Gene Expression." *Phys.org*, July 3, 2023. https://phys.org/news/2023-06-ai-combined-crispr -precisely-gene.html.

Robert, Jason S. 2018. "From Bench to Bedside via . . . The Island of Dr. Moreau?" *Impact Ethics*, November 9, 2018. https://impactethics.ca/2018/11/09/from-bench-to-bedside -via-the-island-of-dr-moreau/.

Sandel, Michael J. 2004. "The Case against Perfection." *Atlantic Monthly*, April 2004. https://www.theatlantic.com/magazine/archive/2004/04/the-case-against-perfection /302927/.

Specter, Michael. 2015. "The Gene Hackers." *New Yorker*, November 8, 2015. https://www .newyorker.com/magazine/2015/11/16/the-gene-hackers.

Wallace-Wells, David. 2023. "Suddenly, It Looks Like We're in a Golden Age for Medicine." *New York Times*, June 23, 2023. https://www.nytimes.com/2023/06/23/magazine /golden-age-medicine-biomedical-innovation.html.

Yong, Ed. 2018. "A Controversial Virus Study Reveals a Critical Flaw in How Science Is Done." *Atlantic Monthly*, October 4, 2018. https://www.theatlantic.com/science /archive/2018/10/horsepox-smallpox-virus-science-ethics-debate/572200/.

I OVERVIEW

The Era of CRISPR

1

CRISPR

Challenges Posed by a Dual-Use Technology

Rachel M. West and
Gigi Kwik Gronvall

What Is CRISPR?

CRISPR, a recently developed gene-editing tool, has become synonymous with rapid biological advancement. While gene editing has been performed in life sciences research for decades, genetic engineering with CRISPR is much more straightforward, faster, and less expensive—allowing the technology to be rapidly democratized. CRISPR was built on a natural mechanism, the method by which bacteria resist infections from viruses called *bacteriophages*. Once infected, bacteria may recognize specific genetic sequences of the invading bacteriophage and chop its genetic material into pieces. This bacterial immune response, discovered through basic research, has been exploited by scientists to develop a gene-editing tool that can selectively find, cut, and replace specific sections of DNA. The work began in 2011–2013 in advanced research laboratories at the University of Vienna, Vilnius University of Lithuania, UC Berkeley, and MIT, but the use and refinement of CRISPR has since expanded across the world to university research laboratories, start-up biotechnology companies, community laboratories, and even DIY (do-it-yourself) Bio science kits.

As with many other powerful biotechnology advances, CRISPR raises dual-use concerns that, in addition to its numerous benefits, it also has the potential to be misused and could lower technical barriers to biological weapons development. Specifically, CRISPR could allow a nefarious actor to edit an existing pathogen to make it more damaging, edit a nonpathogenic organism to incorporate pathogen genes and traits, and even, theoretically, synthesize a novel pathogen. Given CRISPR's affordability, ease of use, and widespread availability, the potential for misuse increases, not only by a malicious actor but also by accident.

This essay describes the expansion and refinement of CRISPR as a genetic-engineering tool; details the consequences of its dual-use potential for benefit and serious harm, including potential biosecurity concerns; and recommends steps governments and scientists may take to reduce biosecurity risks while technology developments proceed. It is not possible to fully eliminate risks from the misuse of biotechnologies, including CRISPR, but steps can be taken to increase safety and security while allowing this powerful technology to remain widely available for beneficial purposes.

How CRISPR Evolved as a Gene-Editing Tool

CRISPR is now a simple, robust, and efficient tool to perform genetic engineering in laboratories all over the world, but it started out as a natural immune mechanism found in bacteria. It helped bacteria resist the incorporation of foreign DNA from either viral bacteriophage threats or conjugation, a process in which one bacterium transfers genetic material to another through direct contact. CRISPR stands for clustered regularly interspaced short palindromic repeats, the repeating genetic sequences found in bacteria and discovered to be pieces of foreign DNA. Though the system requires multiple components, it will be referred to as CRISPR for simplicity throughout this discussion.

This bacterial "immune mechanism" is not as intricate and complex as human immune systems; it is more accurate to think of it as bacterial "memory." Prior exposure to foreign DNA arms bacteria to use CRISPR to prevent similar foreign DNA incorporation in the future. The short palindromic repeats have spacers between them, which are unique sequences that can be shuffled by bacteria over time. These sequences are often small pieces of foreign DNA that a microbe would have previously encountered. Together with the *Cas* enzymes of the CRISPR system, these spacers allow the bacteria to recognize nonself DNA and remove it before it is incorporated into the genome. Hence, the bacteria can have "memory" of a previous infiltration, which helps it respond to any future attempts.

From the laboratories of Doudna (UC Berkeley), Charpentier (Max Planck Institute), Zhang (MIT), and Church (Harvard), CRISPR quickly grew from its natural role in bacteria to a tool for specifically nicking double-stranded DNA and then directing repair via a guide RNA encoded within the plasmid system. CRISPR, as a gene-editing tool, uses a predesigned spacer sequence of DNA and an endonuclease enzyme. While a variety of enzymes have been used, the most developed is *Cas* (Beumer et al. 2023). The system works as follows: a CRISPR sequence (the designed spacer sequence) is transcribed, forming a guide RNA (gRNA) which directs the *Cas* enzyme to the site of interest in the DNA where there is homology. Then, *Cas* makes a cut in the DNA. From there, a known DNA sequence can be inserted as directed by the gRNA, or an additional sequence included, or a process called *nonhomologous end joining* will simply insert random mutations at the cut site. Thus, the CRISPR system can be used to produce a simple mutation to disrupt a gene or a directed mutation, where the scientist provides the desired sequence to be inserted. These mutations can be used in sequence to slowly insert an entire gene into an organism's genome, or a hybrid version of a gene, if the scientist desires.

As a genetic-engineering tool, CRISPR has been used in laboratory research on many varieties of microorganisms, plants, and animals, including mice, goats, and pigs (Wang et al. 2015; Zheng et al. 2017). This has also allowed scientists to tailor model organisms, such as mice, to better represent human diseases.

Funding for CRISPR-related work has rapidly escalated. In 2011, the US National Institutes of Health (NIH) awarded about $5 million for CRISPR-related projects. Seven years later, that funding increased to over $1 billion, including from private donors, with over 12,000 related publications (Congressional Research Service 2018). It has also been used in the creation of diagnostic tools for SARS-CoV-2 (Broughton et al. 2020; Joung et al. 2020). The ease of use and investment in CRISPR technologies ensures its role as a key genetic-engineering tool in the academic and private sectors for years to come.

CRISPR's Expansion to Eukaryotic Cells

Though the Charpentier, Doudna, and Šikšnys laboratories worked extensively on biomolecular characterization of CRISPR, in 2013, the Zhang laboratory at the Broad Institute of MIT and the Church lab of Harvard University were the first to demonstrate its utility in organisms other than bacteria (Broad Institute n.d.). Their innovation centered on using eukaryotic cells, such as cells found in plants and animals, rather than strictly using bacteria or fungi, such as yeast. Before CRISPR, gene editing in eukaryotic cells could be time intensive and complicated.

The variety of endonucleases available has also advanced as CRISPR has been refined as a gene-editing tool (Wang, La Russa, and Qi 2016). The most used system, CRISPR/*Cas9*, is a Class 2. Combined with modification enzymes, this system may be useful to treat rare genetic diseases. Work is now underway on a mouse model with Fragile X syndrome (Liu et al. 2018). In Fragile X, expression of the gene *FMR1* is too low because of DNA methylation modifications. Using CRISPR, DNA methylation can be reduced, making the gene more accessible for transcription and translation and, thus, reducing symptoms of Fragile X.

CRISPR may also be combined with other elements, such as nonpathogenic viruses, to help "carry" the elements to the right place. This is underway in early research to treat neurodegenerative diseases (Gaj and Perez-Pinera 2018). CRISPR/*Cas9* may also be pooled with multiple guide RNAs, allowing the editing of multiple genes in one step. This pool of guide RNAs allows the *Cas* enzyme, which cuts the DNA, to be guided to many distinct parts of the genome. Using pooled guide RNAs to target many genes at once can also be valuable to understand systemic effects, such as impacts on metabolism or response to therapy (Wang, La Russa, and Qi 2016). Further, the CRISPR/*Cas9* system has been modified to become inducible by miRNA (microRNA) to allow for temporal or tissue-specific activity that may be useful for some therapies (Hirosawa et al. 2017).

CRISPR is relatively easy to use, particularly compared to previous methods of gene editing, and it has been demonstrated to cause few off-target effects (where edits are made in the wrong place). This makes it an attractive tool for genetic modification on a larger scale. Nevertheless, the few off-target effects can have major impacts on an organism (O'Geen et al. 2015).

Off-target effects are not exactly like side effects, because they are potentially more damaging. As George Mason University scholars Ben Ouagrham-Gormley and Popescu (2018) explain, these off-target mutations will remain even after CRISPR/*Cas9* systems are removed; in contrast, if a patient stops taking a drug that produces negative side effects, the side effects typically cease. As researcher and past Doudna laboratory member Kyle Watters (2018) asserts, these off-target effects are also of concern in developing CRISPR tools for the biosecurity community, because they can theoretically lead to worsening of disease or fatal outcomes. It can be difficult to measure the potential number of off-target mutations, though methods to find potential error-prone regions are becoming available (Tsai et al. 2017). Using new tools to reduce the risk of off-target effects is promising, but more research is needed to shift the balance in favor of clean, efficient targeting of the gene of interest.

Applications for CRISPR

The uses for CRISPR are diverse and growing, from domestication of crops to antibacterial therapeutics (Greene 2018; Khan et al. 2019). One reason CRISPR is so revolutionary is that scientists can quickly and efficiently tailor the system to target gene(s) of interest, regardless of organism type. It is broadly useful. A gene of interest could be one that has positive impacts on an organism, and scientists may use CRISPR to transfer it to other similar organisms. A gene that has negative impacts could be mutated or deleted. CRISPR may be used in agricultural settings to add desirable traits for improved breeding or hardiness to help domesticate a potentially useful crop. Recent studies have identified, by RNAi silencing, the major allergen gene of peanuts: *Ara h 2*. Silencing this gene appears to reduce the allergenicity of the protein (Dodo 2008). Such a gene could be targeted by CRISPR, and CRISPR plasmids targeting these proteins are currently for sale by Santa Cruz Biotechnologies (Santa Cruz Biotechnology 2019). In bacteria, CRISPR could target antibiotic-resistant genes to restore antibiotic susceptibility and improve treatment options and outcomes.

CRISPR also appears to be an attractive tool for barcoding a set of cells, such as in a tumor, to better understand their development. These barcodes are analogous to trackers in that any cell with a barcode can be sourced to that tumor. This allows the detection of metastasis—the spread of cancer throughout the body. While scientists know that spread and establishment of several tumors in different tissues depends on certain tumor cells escaping, until recently, it was difficult to understand how or why. CRISPR barcoding can help trace how tumor cells travel.

Gene drives are another potential application for CRISPR. A gene drive propagates (or drives) genes into offspring at a higher inheritance rate than would be expected in nature, in what is referred to as *super-Mendelian inheritance*. Offspring not only inherit the modified gene, but they inherit the CRISPR system as well (Synthego 2018). The use of a gene drive should lead to more organisms with the gene (or disrupted gene) of interest—instead of 50% of offspring, as expected in Mendelian inheritance, gene drives can approach 100% inheritance. Gene drive systems have been proposed for pest management, improvement of crop yields, and manipulation of vector populations to reduce the spread of diseases such as malaria. Gene drives proposed for pest management, especially agricultural pests, reduce the fitness of an organism or give it a fatal trait, such as the transformer (*tra*) gene that is essential for female development in the new world screwworm. This pest can be fatal in mammalian hosts, including agricultural livestock, and remains a major agricultural issue, despite previous efforts to use sterile male insect release to reduce its population. Targeting a gene specific to females using a gene drive, according to mathematical models, could suppress the population of new world screwworm in areas where sterile male release is not effective (Scott et al. 2017). The CRISPR gene drive inserts a mutation that has negative effects on an organism and would typically not be advantageous or passed to future generations. With the gene drive, this mutation can persist and reduce the population into the future (McFarlane et al. 2018).

Gene drives have also been proposed to reduce the burden of vector-borne diseases such as malaria, which had 219 million cases and over 435,000 deaths in 2017 alone (World Health Organization [WHO] 2018). While antimalarial drugs and insect control efforts have made a substantial impact on malaria burden over the past decade, the decline has stagnated in recent years. Consequently, novel control efforts are needed, and a gene drive may be another weapon in the fight against malaria.

A gene drive can be used to either suppress the mosquito population—specifically of *Anopheles gambiae*, the primary vector for malaria transmission in Africa—or modify the vector to reduce its capacity to spread malaria. One such gene drive target is a gene called *AGAP005958* that normally confers fertility in female mosquitoes. Identified by the Crisanti Lab in 2016, this gene is expressed in the ovaries, and when there is a complete homozygous deletion or disruption, the females fail to lay eggs. Targeting this gene led to sterility in female mosquitoes, which could be a cost-effective way to suppress large mosquito populations in the wild (Hammond et al. 2016; Hammond and Galizi 2017). Using a gene drive, in addition to other malaria control efforts, could affect the public health of millions without needing extensive human interventions; once the gene drive begins to spread, mosquito mating will maintain the drive. Controversy over the outcome of using gene drives to eliminate malaria-carrying mosquitoes rests on the unpredictability of the full ecological consequences.

DIY Bio and Community Laboratories

CRISPR is not limited to traditional laboratory contexts with academically trained scientists or in the field as a gene drive. Community laboratories and DIY Bio enthusiasts (also called *biohackers*) are using the technology, in many cases to make biological science more accessible for those not in traditional science careers. DIY Bio, a citizen science movement that aims to put science in the hands of the public, is an excellent example of a novel market for CRISPR/*Cas9*. DIY Bio laboratories traditionally operate under low biosafety containment requirements (biosafety level 1 or 2 out of 4, with 4 being the highest level of biocontainment, requiring engineered controls and air supplies) primarily working with nonpathogenic organisms. These are about the same levels of containment that would be seen in a high school laboratory. CRISPR/*Cas9* provides a low-cost method to produce rapid genetic modifications and a rich learning experience about genetics. Kits available online appear to actively market to those in DIY Bio spaces, sometimes disparaging the "traditional" science laboratories from which this technology was developed. One of the more well-known companies is Odin, founded by Jo Zayner, a controversial proponent of citizen science. Zayner's biotechnology supply company has expanded in recent years to include kits for the genetic manipulation of a wide variety of organisms, including plants and animals. One kit, sold on the site for $299 (at the time of this writing), is for genetic modification of tree frogs, with a CRISPR insertion that increases expression of a growth hormone and consequently increases the frogs' size (The Odin 2019).

While these kits and laboratories may provide the public with rich learning experiences in science (and potentially a space for biotechnology entrepreneurs), there is potential for misuse—though at this time, this misuse is related to self-harm. In 2017, the FDA issued a warning against "self-administered gene therapy" after biohackers were attempting to use CRISPR kits on themselves (FDA 2017). The creator of the Odin Kit, Jo Zayner, attempted to inject himself with a CRISPR construct to increase his musculature (Lee 2017). Specifically, he was attempting to target and knock out his myostatin gene (*MSTN*), which would typically inhibit growth of myoblasts, or muscle cells (Zayner 2018). In goats, myostatin knockouts have been demonstrated to alter metabolism and increase muscle mass (He et al. 2018). Another biohacker, Aaron Traywick, injected himself with CRISPR targeting herpesvirus at a Facebook-broadcast event (Mullin 2018). Traywick himself suffered from herpes infection, but he was trying to demonstrate how valuable CRISPR could be as a treatment for common human diseases. Nevertheless, biohackers' attempts to use their own bodies to demonstrate efficacy of a product seems a dangerous precedent for future developers, and one the FDA has publicly decried.

CRISPR as a Potential Biosecurity Hazard

The potential for CRISPR to revolutionize genetic engineering also raises concerns that it could increase biosecurity threats by lowering barriers for biological weapon development. CRISPR may be misused to create increased-virulence pathogens, neurotoxins, and even de novo organisms (DiEuliis, Berger, and Gronvall 2017; DiEuliis and Giordano 2017). A de novo organism would be a completely synthetic organism, though it may have the same genome as an existing pathogen, like smallpox. Synthesizing a completely novel organism is theoretically possible, but it is likely to require extensive training, funding, and time for research and development, which is less possible for some types of actors (Gibson et al. 2010).

A 2018 National Academies of Sciences, Engineering, and Medicine (NASEM) study, "Biodefense in the Age of Synthetic Biology," was undertaken to develop guidance on evaluating biosecurity risks associated with new biotechnology. Among their findings, they categorized potential risks by relative concern. Among the highest of concern is re-creating known pathogens, such as smallpox, while creating novel pathogens is a lower risk (NASEM 2018). CRISPR could allow for rapid, efficient editing of a pathogen to possess the virulence factors of another for the re-creation of a known pathogen whose genome is published. Given these concerns, the use of CRISPR warrants recognition as a potential biosecurity threat if misused.

While not exactly a biosecurity risk, the accessibility of a powerful genetic engineering tool has already led to ethical challenges, such as Chinese scientist He Jiankui's engineering of human embryonic genomes (Cyranoski 2018). His work violated a long-standing norm prohibiting genetic modification of the human germline, which allows modifications to pass on to future generations. The WHO Expert Advisory Committee on Developing Global Standards for Governance and Oversight of Human Genome Editing is an international effort to monitor human genome–editing efforts (WHO 2019). In 2020, the US National Academies also recommended that edited human embryos should not yet be used to create a pregnancy (NASEM 2020).

Gene drives have raised public concern that the changes may be difficult to reverse and may have unintended consequences (Noble et al. 2018). Either by intentional or accidental misuse, CRISPR-based gene drives could have dramatic effects on indigenous species in addition to crops. Further, CRISPR has been suggested as a method to control invasive species (Callaway 2018). Yet engineered organisms may, themselves, become invasive if improperly introduced into the field. It is important to identify the knowledge gaps that complicate efforts to control CRISPR-modified species, because there may be unintended effects, such as altering gene flow within a population (Moro et al. 2018). While many studies have attempted to model gene drives within populations, this cannot completely encompass a complex, dynamic ecosystem (Hayes et al. 2018). For instance, models have indicated that eliminating *Anopheles gambiae* mosquitoes, the vector of the human malaria parasite in Africa, would not have any large effects on the ecosystem (Collins et al. 2019). It is exceedingly difficult to verify this model, because one cannot completely wipe out a species of mosquito to "test" its impact.

Attempts to use gene drives to decimate crops or affect local resources could present a biosecurity threat with a wide range of consequences. Keeping this in mind, the Defense Advanced Research Products Agency (DARPA 2017) created the Safe Genes Project to not only address potential issues in gene drive technology and biosecurity, but to also promote defensive research to create countermeasures.

Amateur and citizen scientist use of CRISPR also presents complications for biosecurity. While many biosecurity issues, such as de novo synthesis of a pathogenic virus, may take an experienced scientist with access to a variety of reagents, CRISPR kits are more widely available to the public. They are intended primarily for learning using nonpathogenic organisms, but they have been used by individuals to attempt (unsuccessfully) to use CRISPR to modify their own genomes. Although misguided, these attempts demonstrate that this can and will happen. Intentional, though inexperienced, actors may use CRISPR to alter existing microorganisms to increase their pathogenicity or to create chimera organisms (Zhang 2018). The democratization of these kits may be an exciting advance for science education, and there is always the potential that biotechnology entrepreneurs will get their start this way, but it does present challenges to traditional biosafety training methods within traditional research institutions. Many DIY Bio laboratories, including Baltimore Underground Science Space (BUGSS 2019), have their own biosafety officers. This allows all members to readily ask questions, prepare their experiments safely, and ensure that all experiments are at the correct biosafety level. Further, specialized FBI agents—WMD Coordinators—often work with DIY Bio laboratories to help explain and enforce laws regarding biosafety (Keulartz and van den Belt 2016). There will always be the possibility of a malicious actor who persists in the DIY Bio space and attempts to misuse technologies such as CRISPR, but the

DIY Bio community seeks to keep their work transparent and is proactive in their commitment to biosafety.

Recommendations to Fortify Biosecurity in the Age of CRISPR

CRISPR is considered a dual-use technology because it can have an array of benefits to science, medicine, and public health and it has the potential to be used maliciously. CRISPR joins an extensive list of powerful biotechnology tools that lower barriers toward biological weapons development and from which it is impossible to eliminate the risks of misuse. There are no "total" solutions that can reduce the risks of misuse to zero. Even if a nation were to outlaw CRISPR, it is already used all over the world. Work in biotechnology and genetic engineering will continue apace in multiple nations; medical countermeasures such as vaccines and drugs will require the use of these technologies as well as the means to detect and attribute misuse. In addition, because the governance of the technologies at the forefront will be enacted by the scientific leaders of the technologies, countries without representation in that group may miss out on their opportunities to shape the rules of the road. While there are no total solutions, partial solutions are possible that can deter nefarious actors, increase the likelihood of detecting and attributing misuse, and limit accidental misuse—while preserving the enormous benefits that CRISPR can bring.

One example of a partial solution is the international governance effort led by the WHO (2014) on the testing and release of genetically modified mosquitoes, including the use of gene drives, for the purpose of malaria control. They provide a framework for how gene drives should be responsibly used and the safety testing required before genetically modified mosquitoes may be released into the wild.

Another successful partial solution is preventing an ill-intended actor from buying the genetic material for a pathogen from a company so it may be made from scratch in a laboratory. In 2010, the US Department of Health and Human Services (HHS) published the *Screening Framework Guidance for Providers of Synthetic Double-Stranded DNA*, which outlined how gene synthesis companies should screen customers and their orders for possible misuse and what they should do if they get a "hit" on a regulated pathogen. Since then, most gene synthesis companies internationally have adopted similar guidelines for their own screening algorithms to actively screen orders (Battelle 2019). There is now a law in California that would make screening a requirement (Molteni 2021). These partial solutions cannot prevent all forms of bioterrorism. It is certainly possible for a rogue actor to acquire pathogens from a variety of places, not just through a gene synthesis company, but these actions raise barriers to misuse.

Governments Should Enhance Scientists' Ability to Self-Govern by Giving Them the Control and Authority that Can Facilitate Self-Governance

Many partial solutions that may increase biosecurity are technical, specific, and emerge as scientific research advances. Therefore, it is critical that scientists pursue self-governance and that governments support them by giving them the tools to develop rules that may become regulations. Scientists are the most familiar with the technological limits and possibilities of the biotechnologies they are developing, and biosecurity considerations should be added to their concerns. Of course, self-governance cannot prevent all misuse, but no other system of governance can, either. Some examples of self-governance include the WHO gene-editing group that will determine the rules for human gene editing, publishing requirements in scientific journals to use institutional review boards and institutional biosafety committees in protocols, regulations that clearly outline what is not allowed, and consulting with scientists at the leading edge of their fields to better assess what controls may be necessary. At the leading edge of research, often well before regulatory structures can be put into place, decisions on pursuing areas of research or whether a procedure is safe must be made by the scientists themselves, often in consensus groups of experts in related fields and including ethicists.

The February 1975 Asilomar conference on recombinant DNA technologies is often cited as a prime example of scientist self-organization and governance to address the potential risks of emerging technologies. In that case, it was the advent of recombinant DNA technology in the 1970s (Berg 2008). Similarly, in response to the He experiments using CRISPR in germline editing, leading scientists including the creators of CRISPR technology recently called for a moratorium on heritable CRISPR editing in humans (Lander et al. 2019). Unlike Asilomar, the act in question has reportedly been accomplished and not prevented, but its repetition has been avoided, as far as we know. The widespread condemnation is good evidence that other scientists will not proceed down He's path until more comprehensive guidance has been agreed upon as to how gene editing should proceed. Scientists are the "boots on the ground" regarding biotechnology and have the potential to be the best reporters of misuse, though there is a long way to go to make sure they understand these responsibilities. This was clearly lacking in the case of He and the other scientists who knew of his plans (Cohen 2019).

Beyond CRISPR, there is an opportunity for the United States and other governments to inform scientists about their responsibilities to protect their research and powerful biotechnology tools and to increase their knowledge of biosecurity. Given that the most egregious examples of biological weapons development and use are increasingly historical, it is strongly suspected that many leading scientists today are broadly unfamiliar with the history or unaware that the tools and technologies of their trade could be misused. The fact that there is a legally binding treaty prohibiting bioweapons development and use—the Convention on the Prohibition of the Development, Production and Stockpiling of Bacteriological and Toxin Weapons and on their Destruction (BWC)—is not expressly taught to scientists. Most scientific research at US universities and federal institutions is funded by taxpayer money, the National Science Foundation, or the NIH (LeMieux 2017). Therefore, guidelines for laboratory training by a federal agency could be relevant to all those receiving funding.

At the Institutional Level, Biosecurity Training Should Be Provided Alongside Research Ethics Courses

Current PhD trainees often participate in research ethics courses, and these could be augmented with training for those working with CRISPR and other genetic-engineering technologies. If scientists learn the risks and guidelines for biosecurity early in their careers, they will be more likely to carry this throughout their work in academia, industry, or other careers. These training modules can be similar in scope and in time commitment to current training on chemical waste and bloodborne pathogens. The training modules could address current guidelines regarding CRISPR technology, legal limits of research (such as germline editing of human embryos), and suggested actions for those with concerns.

Relevant CRISPR Biosafety Guidelines Should Be Made Public to Encourage Nonacademia and DIY Bio Research Safety

Federal guidance may not extend to DIY Bio communities, to nontraditional scientists who desire to use CRISPR outside of a federally funded source, or internationally. CRISPR technologies are widely available at this point through traditional providers such as ThermoFisher Scientific and nontraditional providers such as Odin Technologies. Because of this, it would be incredibly difficult to attempt to address every possible user of CRISPR technologies. There is a risk of accidental misuse in nontraditional science settings, but if institutions and federal funding sources institute relevant guidelines, DIY Bio communities could use them as a model; this work is already being pursued by nongovernmental sources, funded by the Open Philanthropy Project. This could include promoting safe science techniques and discouraging activities such as storing bacterial samples close to food. The potential benefits of the democratization of CRISPR are significant; this makes genomics and gene editing more accessible for those who may not have access to science courses or who want to challenge themselves beyond the classroom. While it may be less likely that amateur scientists will be discovering the applications of CRISPR, which would require cell cultures and potential clinical trials, the learning opportunities are the main benefit. Keeping STEM accessible and diverse can give rise to new leaders in science and bring the next discoveries and businesses focused on biotechnologies.

Education and training of scientists to properly identify and report potential security issues would allow self-governance that would minimally impede scientific growth and innovation. Proper education has also been suggested to facilitate collaboration between scientists and policymakers, because scientists would better understand the legislative perspective regarding biosecurity (Minehata et al. 2013). Governments should create guidance and regulations that support scientists and give them tools for governance. This has been demonstrated by the NIH with their Guidelines for Research Involving Recombinant or Synthetic Nucleic Acid Molecules, which provides a clear framework. The potential for CRISPR technologies is enormous; it can offer vast improvements in therapeutics (for instance, in curing sickle cell disease) and revolutionize gene-editing studies that can reveal even more about the intricacies of the genome. Further, the availability and accessibility of this technology can inspire amateur scientists and the DIY Bio community to promote STEM education, particularly to underserved communities. With small, incremental partial solutions supporting biosecurity, misuse of CRISPR and other genetic-engineering tools could be minimized so we may all benefit from its enormous promise.

REFERENCES

Baltimore Underground Science Space (BUGSS). 2019. "Lab Resources." Accessed October 3, 2019. https://www.bugssonline.org/lab-resources-2/.

Battelle. 2019. "Twist Bioscience Adopts Battelle's ThreatSEQ DNA Screening Web Service for Advanced Biosecurity." Accessed October 3, 2019. https://www.battelle.org/newsroom/press-releases/press-releases-detail/twist-bioscience-adopts-battelle-threatseqtm-dna-screening-web-service-for-advanced-biosecurity.

Ben Ouagrham-Gormley, Sonia, and Saskia Popescu. 2018. "The Dread and the Awe: Crispr's Inventor Assesses Her Creation." *Bulletin of the Atomic Scientists*, March 8, 2018. https://thebulletin.org/2018/03/the-dread-and-the-awe-crisprs-inventor-assesses-her-creation/.

Berg, Paul. 2008. "Meetings That Changed the World: Asilomar 1975; DNA Modification Secured." *Nature* 455 (7211): 290–91.

Beumer, Joep, Hans Clevers, Amandine Caillaud, Bertrand Cariou, Vassilis G. Gorgoulis, Pia Annette Johansson, Heather Keys, et al. 2023. "CRISPR: Questions and Answers." *STAR Protocols*. https://star-protocols.cell.com/protocols/2555.

Broad Institute. n.d. CRISPR Timeline. Accessed November 20, 2023. https://www.broadinstitute.org/what-broad/areas-focus/project-spotlight/crispr-timeline.

Broughton, James P., Xianding Deng, Guixia Yu, Clare L. Fasching, Venice Servellita, Jasmeet Singh, Miao Xin, et al. 2020. "CRISPR-Cas12-Based Detection of SARS-CoV-2." *Nat Biotechnol* 38 (7): 870–74.

Callaway, Ewen. 2018. "Controversial CRISPR 'Gene Drives' Tested in Mammals for the First Time." *Nature* 559 (7713): 164.

Cohen, Jon. 2019. "The Untold Story of the 'Circle of Trust' behind the World's First Gene-Edited Babies." *Science*, August 1, 2019.

Collins, Catherine Matilda, Jane A. S. Bonds, Megan M. Quinlan, and J. D. Mumford. 2019. "Effects of the Removal or Reduction in Density of the Malaria Mosquito, *Anopheles Gambiae s.l.*, on Interacting Predators and Competitors in Local Ecosystems." *Med Vet Entomol* 33 (1): 1–15.

Congressional Research Service. *Advanced Gene Editing: CRISPR-Cas9*. R44824. 2018. https://crsreports.congress.gov/product/pdf/R/R44824.

Cyranoski, David. 2018. "CRISPR-Baby Scientist Fails to Satisfy Critics." *Nature* 564 (7734): 13–14.

Defense Advanced Research Project Agency (DARPA). 2017. "Building the Safe Genes Toolkit." Accessed October 3, 2019. https://www.darpa.mil/news-events/2017-07-19.

DiEuliis, Diane, Kavita Berger, and Gigi Gronvall. 2017. "Biosecurity Implications for the Synthesis of Horsepox, an Orthopoxvirus." *Health Secur* 15 (6): 629–37.

DiEuliis, Diane, and James Giordano. 2017. "Why Gene Editors like CRISPR/Cas May Be a Game-Changer for Neuroweapons." *Health Secur* 15 (3): 296–302.

Dodo, Hortense W., Koffi N. Konan, Fur C. Chen, Marceline Egnin, and Olga M. Viquez. 2008. "Alleviating Peanut Allergy Using Genetic Engineering: The Silencing of the Immunodominant Allergen Ara H 2 Leads to Its Significant Reduction and a Decrease in Peanut Allergenicity." *Plant Biotechnol J* 6 (2): 135–45.

FDA (Food and Drug Administration). 2017. "Information about Self-Administration of Gene Therapy." https://www.fda.gov/vaccines-blood-biologics/cellular-gene-therapy-products/information-about-self-administration-gene-therapy.

Gaj, Thomas, and Pablo Perez-Pinera. 2018. "The Continuously Evolving CRISPR Barcoding Toolbox." *Genome Biol* 19: 143.

Gibson, Daniel G., John I. Glass, C. Lartigue, Vladimir N. Noskov, Ray-Yuan Chuang, Mikkel A. Algire, Gwynedd A. Benders, et al. 2010. "Creation of a Bacterial Cell Controlled by a Chemically Synthesized Genome." *Science* 329 (5987): 52–56.

Greene, Adrienne C. 2018. "CRISPR-Based Antibacterials: Transforming Bacterial Defense into Offense." *Trends Biotechnol* 36 (2): 127–30.

Hammond, Andrew, and Roberto Galizi. 2017. "Gene Drives to Fight Malaria: Current State and Future Directions." *Pathog Glob Health* 111 (8): 412–23.

Hammond, Andrew, Roberto Galizi, Kyros Kyrou, Alekos Simoni, Carla Siniscalchi, Dimitris Katsanos, Matthew Gribble, et al. 2016. "A CRISPR-Cas9 Gene Drive System Targeting Female Reproduction in the Malaria Mosquito Vector *Anopheles Gambiae*." *Nat Biotechnol* 34 (1): 78–83.

Hayes, Keith R., Geoffrey R. Hosack, Genya V. Dana, Scott D. Foster, Jessica H. Ford, Ron Thresher, Adrien Ickowicz, et al. 2018. "Identifying and Detecting Potentially Adverse Ecological Outcomes Associated with the Release of Gene-Drive Modified Organisms." *J Responsible Innov* 5 (sup1): S139–58.

He, Zhengyi, Ting Zhang, Lei Jiang, Minya Zhou, Daijin Wu, Junyan Mei, and Yong Cheng. 2018. "Use of CRISPR/Cas9 Technology Efficiently Targeted Goat Myostatin through Zygotes Microinjection Resulting in Double-Muscled Phenotype in Goats." *Biosci Rep* 38 (6): BSR20180742.

Hirosawa, Moe, Yoshihiko Fujita, Callum J. C. Parr, Karin Hayashi, Shunnichi Kashida, Akitsu Hotta, Knut Woltjen, and Hirohide Saito. 2017. "Cell-Type-Specific Genome Editing with a MicroRNA-Responsive CRISPR–Cas9 Switch." *Nucleic Acids Res* 45 (13): e118.

Joung, Julia, Alim Ladha, Makoto Saito, Nam-Gyun Kim, Ann E. Woolley, Michael Segel, Robert P. J. Barretto, et al. 2020. "Detection of SARS-CoV-2 with SHERLOCK One-Pot Testing." *N Engl J Med* 383 (15): 1492–94.

Keulartz, Jozef, and Henk van den Belt. 2016. "DIY-Bio—Economic, Epistemological and Ethical Implications and Ambivalences." *Life Sci Soc Policy* 12 (1): 7.

Khan, Muhammad Zuhaib, Syed Shan-e-Ali Zaidi, Imran Amin, and Shahid Mansoor. 2019. "A CRISPR Way for Fast-Forward Crop Domestication." *Trends Plant Sci* 24 (4): 293–96.

Lander, Eric S., Françoise Baylis, Feng Zhang, Emmanuelle Charpentier, Paul Berg, Catherine Bourgain, Bärbel Friedrich, et al. 2019. "Adopt a Moratorium on Heritable Genome Editing." *Nature* 567 (7747): 165–68.

Lee, Stephanie M. 2017. "This Guy Says He's the First Person to Attempt Editing His DNA with CRISPR." *BuzzFeed News*, October 14, 2017. https://www.buzzfeednews.com/article/stephaniemlee/this-biohacker-wants-to-edit-his-own-dna#.evELlvD9p.

LeMieux, Julianna. 2017. "How Is Science Funded in the United States?" American Council on Science and Health. Accessed October 3, 2019. https://www.acsh.org/news/2017/02/07/how-science-funded-united-states-10816.

Liu, X. Shawn, Hao Wu, Marine Krzisch, Xuebing Wu, John Graef, Julien Muffat, Denes Hnisz, et al. 2018. "Rescue of Fragile X Syndrome Neurons by DNA Methylation Editing of the FMR1 Gene." *Cell* 172 (5): 979–92.e6.

McFarlane, Gus R., C. Bruce A. Whitelaw, and Simon G. Lillico. 2018. "CRISPR-Based Gene Drives for Pest Control." *Trends Biotechnol* 36 (2): 130–33.

Minehata, Masamichi, Judi Sture, Nariyoshi Shinomiya, and Simon Whitby. 2013. "Implementing Biosecurity Education: Approaches, Resources and Programmes." *Sci Eng Ethics* 19 (4): 1473–86.

Molteni, Megan. 2021. "California Could Be First to Mandate Biosecurity for Mail-Order DNA." *STAT*, May 20, 2021. https://www.statnews.com/2021/05/20/california-could-become-first-state-to-mandate-biosecurity-screening-by-mail-order-dna-companies/.

Moro, Dorian, Margaret Byrne, Malcolm Kennedy, Susan Campbell, and Mark Tizard. 2018. "Identifying Knowledge Gaps for Gene Drive Research to Control Invasive Animal Species: The Next CRISPR Step." *Glob Ecol Conserv* 13: e00363.

Mullin, Emily. 2018. "A Biotech CEO Explains Why He Injected Himself with a DIY Herpes Treatment on Facebook Live." *MIT Technology Review*, February 5, 2018. https://www.technologyreview.com/s/610179/a-biotech-ceo-explains-why-he-injected-himself-with-a-diy-herpes-treatment-live-on-stage/.

National Academies of Sciences, Engineering, and Medicine (NASEM). 2018. *Biodefense in the Age of Synthetic Biology*. Washington, DC: National Academies Press.

National Academies of Sciences, Engineering, and Medicine (NASEM). 2020. "Heritable Human Genome Editing." Accessed July 30, 2021. https://www.nap.edu/catalog/25665/heritable-human-genome-editing.

Noble, Charleston, Ben Adlam, George M. Church, Kevin M. Esvelt, and Martin A. Nowak. 2018. "Current CRISPR Gene Drive Systems Are Likely to Be Highly Invasive in Wild Populations." *ELife* 7: e33423. https://elifesciences.org/articles/33423.

O'Geen, Henriette, Abigail S. Yu, and D. J. Segal. 2015. "How Specific Is CRISPR/Cas9 Really?" *Curr Opin Chem Biol* 29: 72–78.

Santa Cruz Biotechnology. 2019. "ARA CRISPR Plasmids." Accessed October 3, 2019. https://www.scbt.com/browse/ARA-CRISPR-Plasmids/_/N-eosmsa.

Scott, Maxwell J., Fred Gould, Marcé Lorenzen, Nathaniel Grubbs, Owain Edwards, and David O'Brochta. 2017. "Agricultural Production: Assessment of the Potential Use of Cas9-Mediated Gene Drive Systems for Agricultural Pest Control." *J Responsible Innov* 5 (sup1): S98–120.

Synthego. 2018. "Gene Drive Expert Anna Buchman Discusses Controlling Insects with CRISPR." https://www.synthego.com/blog/anna-buchman-gene-drive.

The Odin. 2019. "Frog Genetic Engineering Kit." Accessed October 3, 2019. https://www.the-odin.com/frog-ge-kit/.

Tsai, Shengdar Q., Nhu T. Nguyen, Jose Malagon-Lopez, Ved V. Topkar, Martin J. Aryee, and J. Keith Joung. 2017. "CIRCLE-Seq: A Highly Sensitive In Vitro Screen for Genome-Wide CRISPR–Cas9 Nuclease Off-Targets." *Nat Methods* 14 (6): 607–14.

US Department of Health and Human Services. 2010. *Screening Framework Guidance for Providers of Synthetic Double-Stranded DNA*. Washington, DC. https://www.phe.gov/preparedness/legal/guidance/syndna/documents/syndna-guidance.pdf.

Wang, Haifeng, Marie La Russa, and Lei S. Qi. 2016. "CRISPR/Cas9 in Genome Editing and Beyond." *Annu Rev Biochem* 85 (1): 227–64.

Wang, Xiaolong, Honghao Yu, Anmin Lei, Jiankui Zhou, Wenxian Zeng, Haijing Zhu, Zhiming Dong, et al. 2015. "Generation of Gene-Modified Goats Targeting MSTN and FGF5 via Zygote Injection of CRISPR/Cas9 System." *Sci Rep* 5 (1): 1–9.

Watters, Kyle. 2018. "The CRISPR Revolution: Potential Impacts on Global Health Security." George Mason University. https://mars.gmu.edu/handle/1920/11338?show=full

World Health Organization (WHO). 2014. "The Guidance Framework for Testing Genetically Modified Mosquitoes." https://fctc.who.int/publications/i/item/2014-06-26-the-guidance-framework-for-testing-genetically-modified-mosquitoes.

World Health Organization (WHO). *World Malaria Report*. Geneva: World Health Organization, 2018.

World Health Organization (WHO). *WHO Expert Advisory Committee on Developing Global Standards for Governance and Oversight of Human Genome Editing*. April 16, 2019. https://iris.who.int/bitstream/handle/10665/341017/WHO-SCI-RFH-2019.01-eng.pdf?sequence=1.

Zayner, J. 2018. "True Story: I Injected Myself with a CRISPR Genetic Enhancement." The Antisense. Accessed October 3, 2019. http://theantisense.com/2018/11/13/true-story-i-injected-myself-with-a-crispr-genetic-enhancement/.

Zhang, Sarah. 2018. "A Biohacker Regrets Publicly Injecting Himself with CRISPR." *Atlantic*, February 20, 2018. https://www.theatlantic.com/science/archive/2018/02/biohacking-stunts-crispr/553511/.

Zheng, Qiantao, Jun Lin, Jiaojiao Huang, Hongyong Zhang, Rui Zhang, Xueying Zhang, Chunwei Cao, et al. 2017. "Reconstitution of UCP1 Using CRISPR/Cas9 in the White Adipose Tissue of Pigs Decreases Fat Deposition and Improves Thermogenic Capacity." *Proc Natl Acad Sci U S A* 114 (45): E9474–82.

II   ETHICAL QUESTIONS RAISED BY CRISPR TECHNOLOGY

2

Untangling CRISPR's Twisted Tales

Marcy Darnovsky and
Katie Hasson

The "CRISPR babies" bombshell of November 2018 exploded at the Second International Summit on Human Genome Editing in Hong Kong. Researcher He Jiankui announced the birth of twin girls he dubbed Lulu and Nana, whose genes he had edited as embryos. The revelation was the boiling point in the long-simmering controversy about the prospect of using heritable genome editing (HGE) to alter the genes and traits of future children and generations. He Jiankui's wildly irresponsible experiment dominated the attention of the 500 Summit participants and catapulted debate about HGE out of scientific and science policy circles and into the media, public, and policy mainstreams. A year later, He Jiankui, Lulu, and Nana appeared on nearly every list of the decade's top science stories (Greshko 2019; *Nature* 2019; *New Scientist* 2019; Pappas 2019). He spent three years in jail for practicing medicine without a license and, along with two colleagues, was given hefty fines (Kennedy 2019).

For casual observers, the news stories clearly signaled the seriousness and urgency of the debate about whether HGE should move forward. But the dominant narrative they offered—a "rogue" researcher promptly showered with worldwide denunciations—blurred key fault lines and downplayed the likely social consequences of setting HGE loose on the world.

Unfortunately, many influential accounts were similarly flawed. Since the powerful new gene-editing tool CRISPR came into widespread use after 2012, debates over whether it should be used in human reproduction have proliferated at scientific and science policy meetings; in reports issued by blue-ribbon committees; and in scientific and policy journals, popular science publications, and book-length treatments. The dividing lines in the ongoing controversy are complex, zigzagging among scientists, bioethicists, biotechnology companies, policy experts, public interest advocates, and others. Some prominent individuals and key organizations strongly support prohibiting HGE or urge a moratorium to allow time to seek "broad societal consensus" before policy decisions are reached. Others are doggedly pushing the technology toward clinical trials and fertility clinics.

Our organization, the Center for Genetics and Society, works to bring social justice and human rights perspectives to the public and policy conversations about human genetic and assisted reproductive technologies. Our long-term immersion in the controversy—and our efforts to influence it— have made us keenly aware of the circulating narratives and counternarratives. In this essay, we explore the considerable misrepresentations and distortions that skew public understanding and deform the "sociotechnical imaginary" that shapes our thinking about how gene-editing tools should and should not be used (Jasanoff and Kim 2009).

We support somatic genome editing to treat patients if it is shown to be safe and effective, though we are deeply concerned that therapies be developed and made available in a globally fair and equitable way. By contrast, we believe that HGE would cause dire societal harm and should remain legally off limits, where more than 70 nations and an international treaty have already put it (Baylis et al. 2020). Our goal here is to urge that deliberations about *whether* to proceed be reoriented toward social consequences and significantly broadened in terms of scope, participants, and process.

The CRISPR Babies Aftermath

News reports of He Jiankui's reckless 2018 experiment accurately characterized the swift criticism that followed as "near-universal condemnation" but missed important divergences among those responses. Most notable was the move made by the organizing committee of the gene-editing summit, which had been convened by the science academies of the United States, the United Kingdom, and Hong Kong. Its official concluding statement pivoted directly from denouncing He to proclaiming support for proceeding with heritable genome editing (HGE): "It is time to define a rigorous, responsible translational pathway toward [clinical] trials" (Baltimore et al. 2018).

In an opinion piece titled "Wake-Up Call from Hong Kong," the leadership of the three national academies expressed their support for developing HGE as soon as "criteria and standards are in place," and abandoned the academies' own 2015 standard of "broad *societal* consensus" (Baltimore et al. 2015) in favor of "broad *scientific* consensus" (Dzau, McNutt, and Bai 2018, emphasis added). Shortly thereafter, the US and UK academies convened the International Commission on the Clinical Use of Human Germline Genome Editing, which was tasked with developing "a potential pathway from research to clinical use—if society concludes that heritable human genome editing applications are acceptable" (National Academies of Sciences, Engineering, and Medicine [NASEM] 2019). The He Jiankui incident seemed to be concerning to them mainly because it could stoke public opposition to HGE and undermine scientists' authority over decisions about whether, when, and how it should be used.

Others took note and pointed out that despite gestures toward democratic discussion and governance, proponents were trying to limit the scope and influence of public debate (Baylis and Darnovsky 2019; Hasson and Darnovsky 2018; Hurlbut et al. 2018; Hurlbut, Jasanoff, and Saha 2018). Writing a few months after the Hong Kong summit, two observers noted, "Scientists articulated more concern about maintaining their authority to unilaterally transform human biology than a willingness to have a public debate about the ethics of whether—and under what conditions—such transformation should take place" (Frahm and Doezema 2019).

Numerous prominent scientists and bioethicists were among those skeptical of HGE (Shanks 2019). In early 2019, 18 of them coauthored a comment in *Nature* calling for a moratorium (Lander et al. 2019). Another group of scientists and biotechnology industry figures wrote an open letter to US government agencies in support of a moratorium (American Society of Gene and Cell Therapy 2019). These calls prompted David Baltimore to assert, "To make rules is probably not a good idea" (Saey 2019).

Midway through the same year, Russian scientist Denis Rebrikov announced his plans to proceed with HGE experiments, first saying he would target the same gene He Jiankui had and later shifting the target to a gene for deafness (Cyranoski 2019; Le Page 2019). World Health Organization (WHO) Director-General Dr. Tedros Adhanom Ghebreyesus then issued a statement that it "would be irresponsible at this time" to do so, and that "regulatory authorities in all countries should not allow any further work in this area" (WHO 2019).

The COVID-19 pandemic muted the ongoing debate about HGE, but deliberations and developments have continued. These include influential convenings in 2022 and 2023; reports issued by the WHO (2021) and the US National Academy of Medicine, the US National Academy of Sciences, and the UK Royal Society (2020); and a 2020 Nobel Prize for CRISPR codevelopers Emmanuelle Charpentier and Jennifer Doudna. Robust debate continues well beyond official reports. The past several years have seen numerous widely reviewed new books and documentaries focused significantly on HGE (Cobb 2022; Davies 2020; Greely 2021; Isaacson 2021; Kirksey 2020; Sheehy 2022). He Jiankui's release from prison in 2022 and attempted reentry into the scientific community also generated a flurry of media coverage.

The controversy over HGE continues to unfold. By clarifying key points and concerning gaps in the debate, we hope to make it easier to see what is at stake.

A Closer Look at the Current Debate

A Medical Matter or Parental Preference?

Most discussions of heritable genome editing (HGE) hinge on technical issues and medical trade-offs: whether the procedures can be made safe and effective enough to justify the inevitably remaining risks for any resulting children and their future progeny. But HGE is not a medical matter. Strictly speaking, it would not treat anyone who is alive or prevent anyone from getting sick because it could affect only human beings who have not yet been conceived.

In contrast, somatic gene editing has justifiably revived hopes of medical innovations for existing patients. After several decades of gene therapy disappointments and tragedies, researchers are seeing gratifying successes in using somatic genome editing to treat conditions like sickle cell disease and certain forms of blindness. As with any new medical procedure, safety and efficacy must be rigorously tested. Affordability presents a huge challenge to health equity, with treatment costs currently reaching several million dollars. The key point here for the debate about HGE is that somatic and germline applications are quite different: the former treats patients while the latter alters embryos (Baylis and de Vries 2021; WHO Expert Advisory Committee 2021a, 2021b).

Yet many discussions fail to make this basic distinction clear. To cite one example among many, a 2019 commentary warns that a moratorium on HGE would risk "forestalling life-saving treatments" (Metzl 2019)—a claim that is frequently heard but difficult to understand. Is the author arguing that a moratorium on efforts to create "CRISPR babies" would somehow undercut gene therapies? Is he suggesting that editing the genes of a human embryo should be counted as a "life-saving treatment?" Or does he mean something else entirely?

Statements of this kind lead many people to mistakenly believe that HGE is an exciting new way to save the lives of babies or the only way to prevent the births of children with serious genetic conditions. In fact, anyone who risks passing on genetic disease can avoid doing so by using third-party eggs or sperm. Heterosexual couples who desire full genetic kinship can use preimplantation genetic diagnosis (PGD) to select embryos unaffected by the genetic condition the parents carry. To be sure, selecting embryos—deciding what kinds of people we should welcome into the world—is itself ethically fraught, but making heritable genetic changes would amplify that concern.

The availability of these options makes the medical argument for HGE tenuous at best. But awareness of embryo selection has been low, in part because, until recently, news sources have downplayed or ignored it. Some 15 months after the first use of CRISPR to alter the genes of human embryos in a lab, a review of articles about germline gene editing in major newspapers found that 85% failed to mention PGD (Djoulakian 2016).

Proponents of HGE have advanced several rationales for using it instead of embryo selection. They correctly note that in rare cases, prospective parents would not be able to produce unaffected embryos and could not use PGD. They frequently cite homozygosity for Huntington's disease as an example. But only a minuscule number of people homozygous for Huntington's have been identified in the medical and scientific literature, and sadly, many of them did not live to reproductive age (Lander 2015). Other genetic syndromes that meet these criteria are also exceedingly rare (National Academy of Medicine et al. 2020; Viotti et al. 2019).

Another rationale is that some prospective parents have moral objections to embryo selection, likening it to terminating a pregnancy, and would prefer to "rescue mutant embryos" (Ma et al. 2017). This effort to justify HGE falls flat because PGD would be needed to deselect improperly edited embryos (Ranisch 2019). Other proponents point out, correctly, that multiple IVF cycles would sometimes be required to produce enough embryos for screening, which holds true in general for fertility treatment, not just PGD (Lander et al. 2019).

The crux of the case in favor of HGE thus is not preventing disease, but providing genetically related children to those few people for whom embryo selection would not work. While we can sympathize with parents who place high value on full genetic kinship, we can also recognize this as a personal desire, not a medical need, and weigh this social benefit for a few against the societal risks that HGE would visit on us all (Baylis 2019a; Mills 2019).

Stories about sick babies can be misleading in another way, by causing confusion about the kinds of disorders HGE could even theoretically target. Consider an incident at the first International Summit on Human Gene Editing in 2015. During an open-mic period, a woman named Sarah Gray spoke through tears about watching her baby die shortly after birth. She pleaded passionately for scientists to do whatever it takes to save children like hers—to disregard the risks and "frickin' do it."

Gray's comment was quoted in numerous media reports, including by top science journalists, as a dramatic argument in support of HGE (Begley 2015; Stein 2015; Travis 2015; Yong 2015). But none of these stories noted that HGE was irrelevant to Gray's tragedy. Her baby's condition, anencephaly, has no clear genetic basis. In fact, he had a healthy identical twin; doctors said they found "epigenetic differences in [the babies'] cord blood" (Vitez 2015).

How many people, hearing or reading about this bereaved mother's heart-rending plea, were left with a distorted understanding? Reporters' continued repetition of this story and similar pleas from parents and patients represent but one instance of the hyperbole, exaggerated claims, and misunderstandings prevalent in the current public conversation about HGE (Begley 2019a). By tugging heartstrings with tragic but irrelevant stories and failing to discuss alternatives that would allow parents to avoid passing on adverse genetic conditions, these accounts distort understandings of how, by whom, and to what ends HGE might be used.

Can the Door Be Opened Just a Crack?

Several influential statements about HGE have envisioned approving its use, but in a controlled manner. This was the message of the 2020 report from the US and UK science academies' International Commission, which was explicitly tasked to develop a pathway to the clinic for HGE. The report was widely seen as something of a "do-over" of the science academies' 2017 report, which had generated a list of prerequisites for clinical trials of germline editing that the authors described as so rigorous it might "have the effect of preventing all clinical trials involving germline genome editing" (NASEM 2017, 8). Yet the earlier report's bottom-line recommendation—that such clinical trials be permitted—was widely understood as a green light—including by He Jiankui himself (Lowthorp and Darnovsky 2017; Regalado 2018).

The 2020 report was meant to delve into "practicalities" and stick to "the science" involved in a pathway to the clinic. Its authors were advised to consider ethics and social concerns only where they were "inextricably linked" to the technical and scientific (Begley 2019b). The report specified criteria that must be met for any proposed clinical uses of HGE, including that it be limited to serious monogenic diseases; that no unaffected embryos should be subjected to genetic manipulation; and that prospective parents should have no (or poor) alternatives for having an unaffected, genetically related child (National Academy of Medicine, National Academy of Sciences, and the Royal Society 2020, 100).

Both the 2017 and 2020 report all but ignore the practical feasibility of limiting the use of HGE to specific cases. The 2020 report's suggestion that "an international body" could make key decisions about whether to cross certain thresholds fails to consider how or by whom those decisions would be enforced or how the reliability of these mechanisms might be determined or ensured. Holding this line would be especially challenging in the face of the commercial and marketing pressure that would surely ensue.

In the United States, for example, there is no ready way to control what doctors do once a medical treatment has been approved, because the FDA allows physicians to prescribe "off-label" uses of drugs or devices (Charo 2019). If HGE were to be permitted for a limited set of indications, fertility doctors—who operate in a notoriously underregulated sector—could legally use it for any purpose. While it is possible to imagine both a different regulatory arrangement for off-label use and appropriate regulation over the fertility industry, that is far from the situation in the United States today.

On questions of enforcement, the 2021 WHO committee is much clearer. Their discussion of potential governance mechanisms differs significantly in its focus on the social and political contexts as well as commercial incentives that would complicate global governance, particularly in a situation where countries were allowed to go their own way. The risks of a regulatory patchwork on HGE include exploitation, ethics dumping, and medical or reproductive tourism. While the WHO report stops short of recommending a prohibition or moratorium, the tough questions and serious concerns it raises about responsible global governance of HGE could, if taken seriously, pause the momentum other recent reports seem to set in motion.

In both popular media and "official" reports on HGE, distortions about its potential uses, whom it might help, and how it would be controlled muddy public understanding. Another set of clarifications is needed to address the lack of focus on the social and historical contexts influencing HGE.

Public Policy and Social Consequences

The Current Policy Landscape

One might expect that discussions of heritable genome editing (HGE), especially those organized by well-resourced groups, would accurately depict something as basic as current policies regulating it. But most discussions leave a false impression that such rules are exceedingly rare. Official reports, scholarly accounts (Araki and Ishii 2014; Cavaliere, Devolder, and Giubilini 2019; Daley, Lovell-Badge, and Steffann 2019), and news articles (Kuchler 2020; Mullin 2019) often fail to make clear that numerous laws, regulations, and international agreements are already in place, and glibly assume that global agreement on HGE is unfeasible.

Many such discussions skim over several key but inconvenient facts: the existence of the Oviedo Convention, a binding international treaty prohibiting HGE that has been signed and ratified by 29 countries in the Council of Europe; and prohibitions with the force of law in at least 70 countries, including most nations with significant biotechnological capacities. The 2020 National Academies and Royal Society report, for example, notes several times that HGE is "illegal or otherwise not approved in many countries," but offers no information about which or how many countries. The report mentions the Oviedo Convention once in a footnote. While some proponents have argued that the Oviedo Convention's provision on HGE should be dropped (Sykora and Caplan 2017), the international body has recently reaffirmed it (Council of Europe 2021).

Research we conducted with colleagues identified 75 countries that prohibit HGE (five allow for exceptions) and none that permit it. In other words, among countries with policies on HGE, there is a near consensus that it should be banned. This striking degree of global agreement calls into serious question the assumption in the 2020 National Academies and Royal Society report that countries should and will "go their own way." In our view, any attempt to develop global governance strategies—or even meaningful international cooperation—that fails to acknowledge existing policies in dozens of countries is bound to fall flat.

The United States, unlike many other nations, has had no meaningful legislative debate about HGE. It prohibits HGE with a rider to the federal budget bill, introduced in 2015 and renewed every year since, that bars the FDA from considering clinical trials "in which a human embryo is intentionally created or modified to include a heritable genetic modification" (H.R.2617 - Consolidated Appropriations Act, 2023, Title VII, Sec. 737).

Commercial Dynamics

Assisted reproduction in the United States typically occurs in the private medical sector. Against the background of a tattered and fragmented health-care system, the US fertility industry boasts hundreds of clinics, pulls in annual revenues estimated at $7.9 billion, and operates with minimal oversight (IBISWorld 2023). Fertility procedures are marketed like any other service where extra "add-ons" (many of them untested) are used as enticements. This is the case even in countries such as the United Kingdom, whose national health system offers limited coverage for fertility treatment (Weaver 2018).

Considering the profit potential for the fertility industry, its marketing machinery can be expected to kick into high gear if HGE were permitted. A foretaste is provided by the slick websites of companies selling prenatal cell-free DNA tests and the tests' swift acceptance in routine prenatal care, despite highly misleading claims of accuracy (Kliff and Bhatia 2022) and recommendations for caution by professional medical organizations (Estreich 2019).

One fertility doctor, John Zhang, is open about wanting to develop techniques so "parents can select hair or eye color, or maybe improve their children's IQ." Says Zhang, who gained notoriety when he flouted US law by going to Mexico to produce a baby using a cloning-like technique often referred to as *mitochondrial transfer*, "Everything we do is a step toward designer babies" (Mullin 2017). In fact, Zhang and He Jiankui had plans to open fertility clinics in Hainan, China, that would offer HGE to wealthy prospective parents from around the world (Cohen 2019; Goodyear 2023).

Zhang's designer-baby enthusiasm is unusual in its bluntness, although authors of sober reports also embrace a potential future of human genetic enhancements (Nuffield Council on Bioethics 2018). At the same time, it is increasingly common to hear that we need not worry about designer babies because producing them will always be too technically difficult (Harris 2019; Janssens 2018; Regalado 2019).

It is certainly important to resist overblown claims about the extent to which genes influence traits, especially behavioral and cognitive ones. Yet it is also significant that enhancement enthusiasts are drawing up lists of "protective gene variants" that could give a future child an edge. George Church's list includes genes that confer extra-strong bones, lean muscles, insensitivity to pain, and low odor production (Knoepfler 2015).

Also illustrative is the introduction into fertility clinics of a novel approach to genetic fortune telling: the *polygenic score*. This score is calculated from variations at many genetic loci that supposedly correlate with conditions or traits including diabetes, schizophrenia, and even "educational attainment" as a proxy for IQ (Lee et al. 2018). Companies are already offering to evaluate and "rank" IVF embryos using polygenic scores, despite serious concerns about the underlying science, the proprietary genetic algorithms, and the profound ethical questions this practice raises (Hercher 2021; Lázaro-Muñoz et al. 2021; Turley 2021).

However dubious the claims of genetic prediction or control, the biological reality may turn out to be far less meaningful than the beliefs and dollars invested in such "improvements." Consider the hypothetical wealthy families that would spend hefty sums for genetic upgrades offered by fertility clinics. Whether or not the resulting children's enhancements translated into actual abilities, they would be treated as special by their families, nannies, coaches, and teachers. The belief that they were superior would shape their self-understandings, their relationships, and their lots in life.

Perceptions about genes are indeed powerful, influencing judgments of oneself and others (Matthews et al. 2021) and even altering one's own "physiology, behaviour and subjective experience" (Turnwald et al. 2019, 48). Further, the boosts that children of wealthy families already enjoy due to their socioeconomic privilege could be interpreted as biologically derived and attributed to their genetic upgrades. Would the mere perception that some children were biologically better than others exacerbate social stratification, discrimination, and inequality? The United States has experienced dramatically widening gaps in wealth and income; persistent disparities in access to basic health care, not to mention expensive fertility treatments; and ratcheted-up pressures on and by parents for their children to "get ahead." If HGE were rolled out in fertility clinics, access to it would certainly be stratified, and that disparity would likely be tolerated. Should we risk creating a world of genetic "haves" and "have-nots"?

A Future of Genetic Inequality?

An earlier wave of controversy about altering the genes of future generations swelled at the turn of the millennium, in the aftermath of Dolly the sheep and the run-up to the human genome "map." At that time, both enthusiasts and skeptics were far more likely than participants in today's debates to flag the potential for heritable genetic modification to promote vastly increased inequality. The Princeton molecular geneticist Lee Silver (1997), for example, eagerly anticipated the emergence of a "GenRich" 10% (and dubbed the 90% hoi polloi the "Naturals"). On a global level, Silver (2000) wrote, "the already wide gap between wealthy and poor nations could widen further and further with each generation until all common heritage is gone." The 1997 film *Gattaca*, which remains a touchstone in conversations about genome editing, portrayed a dystopian future of rigid genetic stratification and surveillance.

Well short of a full-blown *Gattaca*-style future, HGE could easily stoke harmful eugenic outcomes and play into the revival of pernicious ideas about the biological bases of race or other socially disfavored traits. This is especially so given the history and legacy of twentieth-century eugenic abuses, the ongoing salience of eugenic temptations, the revival of discredited ideas about race as a biological rather than social and political category, and the new twist on biological determinism in the form of polygenic scores. Any human feature considered less than optimal could be grounds for embryonic editing, from short stature to dark skin or from a low polygenic score for educational attainment to a high one for obesity risk (Khera et al. 2019; Lee et al. 2018). In this context, it would be imprudent to dismiss the potential for a new market-driven eugenics.

People with disabilities were targets of eugenic discrimination, sterilization, and murder as part of the twentieth-century eugenics movement. Since then, disability and social justice advocates have warned about eugenic impulses in prenatal selection practices and in efforts to "fix" or "improve" people with disabilities via genetic modification (Cokley 2017; Estreich 2019; Garland-Thomson 2024; Parens and Asch 1999). Yet some proponents of HGE embrace the term *liberal eugenics* and argue that parents should be allowed, encouraged, or even obligated to "enhance" their future children's traits, starting with the elimination of conditions or traits that might be considered disabilities (Agar 2004; Savulescu and Kahane 2009).

More sober statements about HGE, including the influential reports issued by the US and UK science academies (2017, 2020) and the Nuffield Council (2018), raise these concerns and critiques but diminish their salience by conflating disability with disease and disability rights with patient advocacy. For example, the most recent National Academies report, which focuses on germline editing, uses the words "disability" or "disabilities" 13 times, with

11 of those—85%—as part of the phrase "disease and disability" or a close variant (US National Academy of Medicine, the US National Academy of Sciences, and the UK Royal Society 2020). Disability rights critiques of HGE and the prospect of "velvet eugenics" (Sufian and Garland-Thomson 2021) are acknowledged, but do not persuade these academies and councils against recommending that the technology move forward.

If consequences for people with disabilities are at least mentioned, the dominant discourse largely ignores the impacts of HGE on women (Baylis 2019a). In a commentary on the media coverage of the CRISPR-baby scandal, Rachel Adams (2019) notes that "strikingly absent in the news has been any discussion of where the embryos developed, how the babies came into the world and who will care for them. That is to say, their mother." As a result, Adams says, "It is as if the rogue male scientist is the twins' sole creator."

The absence of women in media discussions on HGE is mirrored in the "big reports," which have little, if anything, to say about the increased risks and pressures assisted reproductive technologies place on women. As an assessment of the Nuffield report points out, "Gene editing of course requires in vitro fertilization (IVF), with its attendant burdens, comparatively low success rates, and possible risks" (Dickenson and Darnovsky 2019). The 2020 National Academies and Royal Society report acknowledges these risks but does so in the context of advocating for HGE as a potential way to reduce the number of cycles women would undergo relative to using PGD. Many women tolerate these adversities in their quests to become mothers (or to sell their eggs or gestational capacities), but that does not mean the risks and burdens of IVF should be missing from considerations of whether to entertain the normalization of HGE.

Outcomes for gene-edited children beyond safety risks must also be explored, though they seldom have been. The experiences of adoptees and the donor conceived demonstrate that children and adults are often intensely interested in their biological origins. Would children who were "designed" feel they were created as products and resent it? Would some parents develop inflated expectations based on the costly outlays they had made? If genetic enhancements become upscale standards, would entirely new familial relationships and predicaments ensue? Considering these possibilities, Bill McKibben (2019a; 2019b) wonders how a firstborn child might experience the birth of a younger sibling with the latest upgrades.

Once the social and political implications of HGE have been recognized, the need to consider a broader range of voices and perspectives becomes clear. On these matters, biomedical expertise is not best suited. Women's health advocates, disabled people and communities, and those working to reduce economic and racial inequities, for example, are uniquely situated to address them. Broad and inclusive public debate should invite and involve these diverse voices and forms of expertise, particularly those coming from organized civil society, where groups and individuals are closely attuned to the social inequities that would influence the use of HGE—and that could be exacerbated by it (Piracés 2018; WHO Expert Advisory Committee 2021a, 2021b).

Opening Up the Conversation

How can we foster constructive conversations about heritable genome editing (HGE) and make sure they are considered in consequential decision-making processes? First, we must broaden our perspective on who should have a say and take a more expansive view of what is at stake. Beyond that, designing forums and other mechanisms for meaningful deliberation by empowered publics and building a political and scientific culture that takes the conversation seriously will require considerable creativity, time, effort, and resources.

Public Engagement that Falls Short

Across the spectrum of views on HGE, there is surprising agreement on the need for public participation and debate. It is widely acknowledged that the issues raised go well beyond scientific laboratories and fertility clinics and warrant democratic deliberation and governance. This recognition was behind the call for "broad societal consensus" in the final statement by the organizers of the 2015 International Summit on Human Gene Editing (Baltimore et al. 2015). But what is meant by public participation can vary, and the elite group that dominates the science academies' work on HGE has been strenuously backpedaling on that early commitment (Baylis and Darnovsky 2019). Many of the models for public engagement that have been employed or proposed fall short by limiting, devaluing, or undermining genuine participation of diverse publics (Andorno et al. 2020). So far, the science academies' leaders have initiated and shaped what passes for public and policy debate on the subject, even while staking a claim for scientific self-regulation of this socially consequential tool (Dzau, McNutt, and Bai 2018; NASEM 2019).

One way scientists assert control is by insisting that discussions be limited to what they assess as technologically possible. This produces a self-fulfilling prophecy in which law and ethics forever "lag" behind the rapid pace of technological developments (Jasanoff 2019). A version of this can be seen in the 2020 National Academies and Royal Society report. Even though the committee's mission to develop a "translational pathway" is framed with the qualifier "should society conclude such applications are acceptable," the report effectively punts responsibility for public participation to the WHO committee. However, this deferral comes packaged with a fully elaborated "translational pathway" for bringing HGE to the clinic. As bioethicist Peter Mills (2020) of the Nuffield Council pointed out in response: "The pathway is a powerful organizing metaphor for technology. The pathway drives traffic to its destination. If you have not asked for a pathway, or if you don't want a pathway, being presented with fully drawn-up plans for a pathway may seem a little presumptuous."

Public participation models focused on "education" can similarly have the effect of narrowing the debate to scientific facts and individual medical risks, reinforcing the assumption that scientists are the appropriate authorities. In this model, because any opposition must stem from lack of knowledge, everyone will agree with going forward once that knowledge deficit is corrected. When public objections do not fit into this frame, they are often rejected as ignorant or emotional. Of course, these assumptions constrict or eliminate public deliberation and influence decision-making (Baylis 2019a).

Opinion polls are sometimes suggested to capture what "the public" thinks about the issue of HGE. But polls tightly limit the range of potential viewpoints respondents can express, reducing the complexities surrounding human genome editing to a series of yes-or-no or multiple-choice questions. They are also extremely sensitive to framing and wording. Typically, survey instruments have not included essential facts about HGE, such as the availability of established alternatives (Neergaard 2018). At best, polls provide a snapshot with uncertain meaning and no opportunity for dialogue.

More interactive public engagement processes can also be problematic. Often, they convene small groups of laypeople to learn about and discuss controversial topics. These time-constrained events begin by teaching participants "the science" from predesigned materials and presentations that, unless carefully prepared, can skew rather than promote understanding. At a recent Australian public engagement exercise on HGE, nearly half the participants expressed afterward that the program had been imbalanced, with an "absence of advocates speaking in opposition to genome editing" (Nicol et al. 2022). Discussion topics and endpoints are often determined in advance. Meaningful mechanisms that would allow the discussion to influence policymaking decisions are seldom provided (Baylis 2019a).

In the United Kingdom, public consultation processes about new human genetic technologies have become routine. Though on their face they can appear impressive, at least some seem carefully managed to produce a predetermined outcome. One such exercise was undertaken in 2013 to gauge public opinions about nuclear genome transfer (so-called "mitochondrial replacement"), a form of cellular engineering that, like HGE, would be passed down through the generations (HFEA 2013). The consultation claimed to find "broad public support" for the procedure in question, but a closer look casts serious doubt on its conduct and conclusions (Cussins and Lowthorp 2018; Cussins and Shanks 2013).

None of these efforts have emphasized the element of shared decision-making that Françoise Baylis (2019a) has argued is the key to public empowerment. Efforts to develop robust forms of public engagement that are up to this task should begin by expanding our understanding of who should be at the table and what issues should be placed on it.

Toward Public Empowerment

Some participants in the "official" deliberations about HGE object to the call for "broad societal consensus," dismissing it as "simply impossible" (Charo 2019, 977). Yet several attempts are already underway to mobilize the creativity, effort, and resources that will be required to engage the public. One influential academic group has established the Global Observatory for Genome Editing to explore key questions about humanity and society in the context of emerging biotechnologies and host the kinds of international and interdisciplinary conversations that might facilitate "broad societal consensus" (Hurlbut et al. 2018; Jasanoff and Hurlbut 2018; Saha et al. 2018). Our organization, the Center for Genetics and Society, has launched a collaborative effort called the Missing Voices Initiative, with participants from a range of social justice advocacy sectors and scholars from the humanities and social sciences. The initiative has organized a robust series of virtual events and is currently developing principles and model policies on HGE based in gender justice and disability rights (Center for Genetics and Society n.d.). These efforts model broad and inclusive participation and demonstrate the starkly different questions and concerns that emerge when conversations foreground social justice perspectives on HGE and its implications for a fair, inclusive, and sustainable future.

Baylis (2019a) envisions "public empowerment" as a process-based alternative to public education or engagement, in which "broad societal consensus" is defined not as unanimity or majority rule, but rather as unity. She elaborates: "Decision making by consensus is about engaged, respectful dialogue and deliberation, where all participants recognize at the outset that knowledge is value laden; that we can and should learn from each other; and that no one should impose his or her will on others" (Baylis 2019b).

These and other proposals are promising but challenging. As the WHO committee cautions, without an investment of "a considerable amount of time and resources" into meaningful public engagement, any efforts risk being seen as "empty public relations or window dressing" (WHO Expert Advisory Committee 2021a, 18).

Conclusion
What's at Stake and Who Is at the Table?

As we work to open and enrich these deliberations, the first step is recognizing how the existing discourse shapes the kinds of public engagement and empowerment that are possible. Hyperbolic claims that link heritable genome editing (HGE) to "life-saving treatments" produce false hopes for parents and patients, train the spotlight on technical rather than social issues, and hand control of the conversation to scientists, doctors, and medical ethicists. While societal risks may be mentioned, they remain unexplored. Despite copious attention to HGE in recent years, there has been no sustained and careful consideration of social and historical contexts, commercial dynamics and pressures, or systemic inequalities and discrimination. The science academies' reports assume that HGE will soon be shown safe and effective; they proceed from there to introduce guidelines for how it should be put into practice. Sorting out the question of whether it should be used at all is left in the dust.

Genuine inclusivity in these deliberations will require challenging the medical framing that favors professional and technical authority, constrains the questions asked, and delimits which risks and benefits are seen as relevant. Importantly, public deliberation requires an open scope that keeps the question of whether to use HGE at all on the table. While the societal concerns raised in this essay ought to be central, part of the goal of genuine public deliberation should be to allow as-yet-unforeseen questions and issues to emerge.

Despite decades of speculation about genetically modified humans, blaring headlines about CRISPR babies, and bans in more than 70 nations, HGE has not found a prominent place on the mainstream political or policy agenda, especially in the United States. But there are signs that this may be changing. A proposed 2019 US Senate resolution (S. Res. 275) suggested that some elected officials are starting to pay attention. The presence among President Joseph R. Biden's top science advisors of individuals who have been active in these discussions suggested that the topic might receive increased attention in this administration.

Civil society has also become increasingly active, particularly in response to He Jiankui's reckless actions (Center for Genetics and Society 2018) and in connection with the Third International Genome Editing Summit (Hasson 2023). It is possible that this activity had some influence on the final statement from the organizers of the 2023 Summit, which stated in part: "Heritable human genome editing *remains unacceptable* at this time. Public discussions and policy debates continue and are important for resolving *whether* this technology should be used" (Lovell-Badge et al. 2023, emphasis added). This tone is quite different from the "let's move forward" call issued in the 2018 Summit statement and seems to indicate at least a temporary retreat by proponents of HGE—one that could allow time for public empowerment and meaningful deliberations by civil society. Perhaps, in a few years, we will see the church basement meetings, webinars, and letter-writing campaigns that characterize other social movements confronting issues with significant scientific and technical components (such as activism focused on AIDS, breast cancer, climate change, and nuclear power).

Reflections on COVID-19

The COVID-19 pandemic invites us to reflect on the inextricable relationship between biomedical science and society. The remarkably rapid development of highly effective vaccines was indeed a triumph of biomedical research. But the pandemic also made clear the value and necessity of other forms of knowledge. Some of its challenges were most effectively addressed by expertise related to the virus and the vaccines (public health, regulatory protocols, health care at all levels, and science communications); others required entirely different forms of knowledge, like those held by essential frontline workers, health equity advocates, and experts in supply-chain logistics.

Another insight from COVID-19 that we can apply to HGE derives from the vast inequalities within and among nations. The SARS-CoV-2 virus and its mutations threaten everyone, but its harms hit selectively. Black and

Latinx communities, for example, have experienced hospitalization and death rates far higher than those in white communities (Holpuch 2021; Williams and Cooper 2020). While part of the disparity in individual outcomes has a biological explanation (notably the greater danger of serious disease in older and immunocompromised people), much of it is determined by socioeconomic factors. They include, importantly, the social determinants of preexisting health conditions like type 2 diabetes and of vaccine uptake, exposures experienced while earning one's living and other activities of daily life, and disability status. That these socially determined differences are so powerfully correlated with the distribution of pandemic harms (and benefits) is unsurprising. It lends weight to the expectation of unjust and discriminatory outcomes in the wake of other sorts of large-scale disruptions, including those that would accompany the introduction of a powerful new reproductive practice involving HGE.

Global inequalities in health care have also been laid bare, especially by the vast chasm in vaccine availability between wealthy and other nations. In North America and Europe, population-wide boosters were administered even as health-care workers and vulnerable groups in most of Asia and Africa remained unprotected. Entreaties that rich countries adjust intellectual property claims so vaccine production could be set up around the world were repeatedly dismissed. This political and moral travesty, which sickened and killed many thousands, is a dramatic demonstration of the truism that "science" is embedded in social relations and power dynamics. Billionaire philanthropist Strive Masiyiwa, serving as the African Union's envoy for vaccine acquisition, asked, "How can I say science has delivered a miracle to people who are dying?" (Milken Institute Staff 2021).

A global pandemic and speculative genetic technologies are distinct problems, but both raise questions about what kind of world we want to live in and what steps we can take now to bring it about. Calls to view the pandemic as a "portal" to a more just future are also warnings about the imperative—and the challenge—of fostering global cooperation before time runs out. As we contemplate the trajectory of HGE, perhaps the most obvious lesson of the pandemic is the basic point that science is inseparable from society, politics, and power—in the questions asked in its name; in its routine practices and extraordinary moments; and in its effects on individuals, communities, society, and the human future.

REFERENCES

Adams, Rachel. 2019. "Gene-edited babies don't grow in test tubes—mothers' roles shouldn't be erased." May 29, 2019. *Conversation*. https://theconversation.com/gene -edited-babies-dont-grow-in-test-tubes-mothers-roles-shouldnt-be-erased-117070.

Agar, Nicholas. 2004. *Liberal Eugenics: In Defence of Human Enhancement*. Hoboken: Wiley-Blackwell.

American Society of Gene and Cell Therapy. 2019. "Scientific Leaders Call for Global Moratorium on Germline Gene Editing." April 24, 2019. https://asgct.org/research/news /april-2019/scientific-leaders-call-for-global-moratorium-on-g.

Andorno, Roberto, Françoise Baylis, Marcy Darnovsky, Donna Dickenson, Hille Haker, Katie Hasson, Leah Lowthorp, et al. 2020. "Geneva Statement on Heritable Human Genome Editing: The Need for Course Correction." *Trends Biotechnol* 38 (4): 351–54.

Araki Motoko, and Tetsuya Ishii. 2014. "International Regulatory Landscape and Integration of Corrective Genome Editing into In Vitro Fertilization." *Reprod Biol Endocrinol* 12 (108). https://doi.org/10.1186/1477-7827-12-108.

Baltimore, David, Françoise Baylis, Paul Berg, George Q. Daley, Jennifer A. Doudna, Eric S. Lander, Robin Lovell-Badge, et al. 2015. *On Human Gene Editing: Statement by the Organizing Committee of the First International Summit on Human Gene Editing*. Washington, DC: National Academies Press. https://www.nationalacademies.org /news/2015/12/on-human-gene-editing-international-summit-statement.

Baltimore, David, R. Alta Charo, George Q. Daley, Jennifer A. Doudna, Kazuto Kato, Jin-Soo Kim, Robin Lovell-Badge, et al. 2018. *On Human Genome Editing II: Statement by the Organizing Committee of the Second International Summit on Human Genome Editing*. Washington, DC: National Academies Press. https://www.nationalacademies.org /news/2018/11/statement-by-the-organizing-committee-of-the-second-international -summit-on-human-genome-editing.

Baylis, Françoise. 2019a. *Altered Inheritance: CRISPR and the Ethics of Human Genome Editing*. Cambridge: Harvard University Press.

Baylis, Françoise. 2019b. "Before Heritable Genome Editing, We Need Slow Science and Dialogue 'within and across Nations.'" *STAT*, September 23, 2019. https://www .statnews.com/2019/09/23/genome-editing-slow-science-dialogue/.

Baylis, Françoise, and Marcy Darnovsky. 2019. "Scientists Disagree about the Ethics and Governance of Human Germline Editing." *Hastings Center*, January 17, 2019. https:// www.thehastingscenter.org/scientists-disagree-ethics-governance-human-germline -genome-editing/.

Baylis, Françoise, Marcy Darnovsky, Katie Hasson, and Timothy M. Krahn. 2020. "Human Germline and Heritable Genome Editing: The Global Policy Landscape." *CRISPR J* 3 (5): 365–77. https://doi.org/10.1089/crispr.2020.0082.

Baylis, Françoise, and Jantina de Vries. 2021. "Equity and Access Need to Be at the Forefront of Innovation in Human Genome Editing." *Conversation*, July 12, 2021. https:// theconversation.com/equity-and-access-need-to-be-at-the-forefront-of-innovation-in -human-genome-editing-161794.

Begley, Sharon. 2015. "Dare We Edit the Human Race? Star Geneticists Wrestle with Their Power." *STAT*, December 2, 2015. https://www.statnews.com/2015/12/02/gene-editing -summit-embryos/.

Begley, Sharon. 2019a. "As Calls Mount to Ban Embryo Editing with CRISPR, Families Hit by Inherited Diseases Say, Not So Fast." *STAT*, April 17, 2019. https://www.statnews.com/2019/04/17/crispr-embryo-editing-ban-opposed-by-families-carrying-inherited-diseases/.

Begley, Sharon. 2019b. "For Rules on Creating 'CRISPR Babies' from Edited Embryos, Scientists Call a Do-Over." *STAT*, August 12, 2019. https://www.statnews.com/2019/08/12/crispr-babies-rules-scientists-call-for-do-over/.

Cavaliere, Giulia, Katrien Devolder, and Alberto Giubilini. 2019. "Regulating Genome Editing: For an Enlightened Democratic Governance." *Camb Q Healthc Ethics* 28 (1): 76–88. https://doi.org/10.1017/S0963180118000403.

Center for Genetics and Society. 2018. "Civil Society Statement to the Organizers of the Second International Summit on Human Genome Editing." November 28, 2018. https://www.geneticsandsociety.org/internal-content/civil-society-statement-organizers-second-international-summit-human-genome.

Center for Genetics and Society. n.d. "Missing Voices Initiative." Accessed March 29, 2023. https://www.geneticsandsociety.org/internal-content/missing-voices-initiative-1.

Charo, R. Alta. 2019. "Rogues and Regulation of Germline Editing." *N Engl J Med* 380 (10): 976–80. https://doi.org/10.1056/NEJMms1817528.

Cobb, Matthew. 2022. *As Gods: A Moral History of the Genetic Age*. New York: Basic Books.

Cohen, Jon. 2019. "The Untold Story of the 'Circle of Trust' behind the World's First Gene-edited Babies." *Science*. August 1, 2019. https://www.science.org/content/article/untold-story-circle-trust-behind-world-s-first-gene-edited-babies.

Cokley, Rebecca. 2017. "Please Don't Edit Me Out." *Washington Post*, August 10, 2017. https://www.washingtonpost.com/opinions/if-we-start-editing-genes-people-like-me-might-not-exist/2017/08/10/e9adf206-7d27-11e7-a669-b400c5c7e1cc_story.html.

Council of Europe. 2021. "Genome Editing Technologies: Some Clarifications but No Revision of the Oviedo Convention." June 4, 2021. https://www.coe.int/en/web/bioethics/-/genome-editing-technologies-some-clarifications-but-no-revision-of-the-provisions-of-the-oviedo-convention.

Cussins, Jessica, and Leah Lowthorp. 2018. "Germline Modification and Policymaking: The Relationship Between Mitochondrial Replacement and Gene Editing." *New Bioeth* 24 (1): 74–94. https://doi.org/10.1080/20502877.2018.1443409.

Cussins, Jessica, and Pete Shanks. 2013. "Broad Public Support for '3-Parent Babies' and Crossing the Human Germline? Not What the Data Say." *Biopolitical Times*, March 21, 2013. https://www.geneticsandsociety.org/biopolitical-times/broad-public-support-3-parent-babies-and-crossing-human-germline-not-what-data.

Cyranoski, David. 2019. "Russian Biologist Plans More CRISPR-Edited Babies." *Nature* 570 (7760): 145–46. https://doi.org/10.1038/d41586-019-01770-x.

Daley, George Q., Robin Lovell-Badge, and Julie Steffann. 2019. "After the Storm—A Responsible Path for Genome Editing." *N Engl J Med* 380: 897–99. https://doi.org/10.1056/NEJMp1900504.

Davies, Kevin. 2020. *Editing Humanity: The CRISPR Revolution and the New Era of Genome Editing*. New York: Pegasus Books.

Dickenson, Donna, and Marcy Darnovsky. 2019. "Did a Permissive Scientific Culture Encourage the 'CRISPR Babies' Experiment?" *Nat Biotechnol* 37: 355–57. https://www.nature.com/articles/s41587-019-0077-3.

Djoulakian, Hasmik. 2016. "Editorial Precision? Snapshot of CRISPR Germline in the News." *Biopolitical Times*, August 1, 2016. https://www.geneticsandsociety.org/biopolitical-times/editorial-precision-snapshot-crispr-germline-news.

Dzau, Victor J., Marcia McNutt, and Chunli Bai. 2018. "Wake-Up Call from Hong Kong." *Science* 362 (6420): 1215. https://doi.org/10.1126/science.aaw3127.

Estreich, George. 2019. *Fables and Futures: Biotechnology, Disability, and the Stories We Tell Ourselves*. Cambridge: MIT Press.

Frahm, Nina, and Tess Doezema. 2019. "Are Scientists' Reactions to 'CRISPR Babies' about Ethics or Self-Governance?" *STAT*, January 28, 2019. https://www.statnews.com/2019/01/28/scientists-reactions-crispr-babies-ethics-self-governance/.

Garland-Thomson, Rosemarie. 2024. "Velvet Eugenics: In the Best Interests of Our Future Children." In *The Promise and Peril of CRISPR*, edited by Neal Baer, 153–71. Baltimore: Johns Hopkins.

Goodyear, Dana. 2023. "The Transformative and Alarming Power of Gene Editing." *New Yorker*. September 2, 2023. https://www.newyorker.com/magazine/2023/09/11/the-transformative-alarming-power-of-gene-editing.

Greely, Henry T. 2021. *CRISPR People: The Science and Ethics of Editing Humans*. Cambridge: MIT Press.

Greshko, Michael. 2019. "These Are the Top 20 Scientific Discoveries of the Decade." *National Geographic*, December 5, 2019. https://www.nationalgeographic.com/science/article/top-20-scientific-discoveries-of-decade-2010s.

Harris, Richard, presenter. 2019. "Why Making A 'Designer Baby' Would Be Easier Said than Done." *All Things Considered*. Aired May 2, 2019, on NPR. https://www.npr.org/sections/health-shots/2019/05/02/719665841/why-making-a-designer-baby-would-be-easier-said-than-done.

Hasson, Katie. 2023. "Dispatch from the Gene Editing Summit." Center for Genetics and Society, March 10, 2023. https://www.geneticsandsociety.org/biopolitical-times/dispatch-gene-editing-summit.

Hasson, Katie, and Marcy Darnovsky. 2018. "Gene-Edited Babies: No One Has the Moral Warrant to Go It Alone." *Guardian*, November 27, 2018.

Hercher, Laura. 2021. "A New Era of Designer Babies May Be Based on Overhyped Science." *Scientific American*, July 12, 2021. https://www.scientificamerican.com/article/a-new-era-of-designer-babies-may-be-based-on-overhyped-science/.

Holpuch, Amanda. 2021. "Covid's Racial Impact in US Clouded by Failure to Collect Race and Ethnicity Data." *Guardian*, July 16, 2021. https://www.theguardian.com/us-news/2021/jul/16/us-covid-coronavirus-race-ethnicity-data.

Human Fertilisation and Embryology Authority (HFEA). 2013. *Medical Frontiers: Debating Mitochondria Replacement, Annex IV: Summary of the 2012 Open Consultation Questionnaire*. London: Dialogue by Design.

Hurlbut, J. Benjamin. 2017. *Experiments in Democracy: Human Embryo Research and the Politics of Bioethics*. New York: Columbia University Press.

Hurlbut, J. Benjamin. 2019. "Human Genome Editing: Ask Whether, Not How." *Nature* 565 (7738): 135. https://doi.org/10.1038/d41586-018-07881-1.

Hurlbut, J. Benjamin, Sheila Jasanoff, and Krishanu Saha. 2018. "The Chinese Gene-Editing Experiment Was an Outrage: The Scientific Community Shares Blame." *Washington Post*, November 29, 2018. https://www.washingtonpost.com/outlook/2018/11/29/chinese-gene-editing-experiment-was-an-outrage-broader-scientific-community-shares-some-blame/.

Hurlbut, J. Benjamin, Sheila Jasanoff, Krishanu Saha, Aziza Ahmed, Anthony Appiah, Elizabeth Bartholet, Françoise Baylis, et al. 2018. "Building Capacity for a Global Genome Editing Observatory: Conceptual Challenges." *Trends Biotechnol* 36 (7): 639–41. https://doi.org/10.1016/j.tibtech.2018.04.009.

IBISWorld. 2023. "Fertility Clinics in the US—Market Size, Industry Analysis, Trends and Forecasts (2023–2028)." *IBISWorld*. Updated September 2023. https://www.ibisworld.com/united-states/market-research-reports/fertility-clinics-industry/.

Isaacson, Walter. 2021. *The Code Breaker: Jennifer Doudna, Gene Editing, and the Future of the Human Race*. New York: Simon & Schuster.

Janssens, A. Cecile J. W. 2018. "Those Designer Babies Everyone Is Freaking Out about—It's Not Likely to Happen." *Conversation*, December 10, 2018. https://theconversation.com/those-designer-babies-everyone-is-freaking-out-about-its-not-likely-to-happen-103079.

Jasanoff, Sheila. 2019. *Can Science Make Sense of Life?* Cambridge: Polity Press.

Jasanoff, Sheila, and J. Benjamin Hurlbut. 2018. "A Global Observatory for Gene Editing." *Nature* 555: 435–37. https://doi.org/10.1038/d41586-018-03270-w.

Jasanoff, Sheila, and Sang-Hyun Kim. 2009. "Containing the Atom: Sociotechnical Imaginaries and Nuclear Power in the United States and South Korea." *Minerva* 47 (2): 119–46. https://www.jstor.org/stable/41821489?seq=1#page_scan_tab_contents.

Kennedy, Merrit. 2019. "Chinese Researcher Who Created Gene-Edited Babies Sentenced to 3 Years in Prison." *NPR News*, December 30, 2019. https://www.npr.org/2019/12/30/792340177/chinese-researcher-who-created-gene-edited-babies-sentenced-to-3-years-in-prison.

Khera, Amit V., Mark Chaffin, Kaitlin H. Wade, Sohail Zahid, Joseph Brancale, Rui Xia, Marina Distefano, et al. 2019. "Polygenic Prediction of Weight and Obesity Trajectories from Birth to Adulthood." *Cell* 177 (3): 587–96. https://doi.org/10.1016/j.cell.2019.03.028.

Kirksey, Eben. 2020. *The Mutant Project: Inside the Global Race to Genetically Modify Humans*. New York: St. Martin's Press.

Kliff, Sarah, and Aatish Bhatia. 2022. "When They Warn of Rare Disorders, These Prenatal Tests Are Usually Wrong." *New York Times*, January 1, 2022. https://www.nytimes.com/2022/01/01/upshot/pregnancy-birth-genetic-testing.html.

Knoepfler, Paul. 2015. "A Conversation with George Church on Genomics and Germline Human Genetic Modification." *Niche*, March 9, 2015. https://ipscell.com/2015/03/georgechurchinterview/.

Kuchler, Hannah. 2020. "Jennifer Doudna, CRISPR Scientist, on the Ethics of Editing Humans." *Financial Times*, January 31, 2020. https://www.ft.com/content/6d063e48-4359-11ea-abea-0c7a29cd66fe.

Lander, Eric S. 2015. "Brave New Genome." *N Engl J Med* 373: 5–8. https://www.nejm.org/doi/full/10.1056/NEJMp1506446#t=article.

Lander, Eric S., Françoise Baylis, Feng Zhang, Emmanuelle Charpentier, Paul Berg, Catherine Bourgain, Bärbel Friedrich, et al. 2019. "Adopt a Moratorium on Heritable Genome Editing." *Nature* 567 (7747): 165–68. https://doi.org/10.1038/d41586-019-00726-5.

Lázaro-Muñoz, Gabriel, Stacey Pereira, Shai Carmi, and Todd Lencz. 2021. "Screening Embryos for Polygenic Conditions and Traits: Ethical Considerations for an Emerging Technology." *Genet Med* 23 (3): 432–34. https://doi.org/0.1038/s41436-020-01019-3.

Lee, James J., Robbee Wedow, Aysu Okbay, Edward Kong, Omeed Maghzian, Meghan Zacher, Tuan Anh Nguyen-Viet, et al. 2018. "Gene Discovery and Polygenic Prediction from a Genome-Wide Association Study of Educational Attainment in 1.1 Million Individuals." *Nat Genet* 50 (8): 1112–21. https://www.nature.com/articles/s41588-018-0147-3.

Le Page, Michael. 2019. "Five Couples Lined Up for CRISPR Babies to Avoid Deafness." *New Scientist*, July 4, 2019. https://www.newscientist.com/article/2208777-exclusive-five-couples-lined-up-for-crispr-babies-to-avoid-deafness/.

Lovell-Badge, Robin, David Baltimore, Françoise Baylis, Ewan Birney, R. Alta Charo, George Daley, Javier Guzman, et al. 2023. "Statement from the Organising Committee of the Third International Summit on Human Genome Editing." https://royalsociety.org/news/2023/03/statement-third-international-summit-human-genome-editing/.

Lowthorp, Leah, and Marcy Darnovsky. 2017. "Reproductive Genome Editing and the U.S. National Academies Report: Knocking on a Closed Door or Throwing It Wide Open?" *Bioethica Forum* 10 (2): 65–67. http://www.bioethica-forum.ch/docs/17_2/06_Lowthorp_BF10_02_2017.pdf.

Ma, Hong, Nuria Marti-Gutierrez, Sang-Wook Park, Jun Wu, Yeonmi Lee, Keiichiro Suzuki, Amy Koski, et al. 2017. "Correction of a Pathogenic Gene Mutation in Human Embryos." *Nature* 548 (7668): 413–19. https://www.nature.com/articles/nature23305.

Matthews, Lucas J., Matthew S. Lebowitz, Ruth Ottman, and Paul S. Appelbaum. 2021. "Pygmalion in the Genes? On the Potentially Negative Impacts of Polygenic Scores for Educational Attainment." *Social Psychology of Education* 24 (3): 789–808. https://doi.org/10.1007/s11218-021-09632-z.

McKibben, Bill. 2019a. *Falter: Has the Human Game Begun to Play Itself Out?* New York: Henry Holt and Company.

McKibben, Bill. 2019b. "Climate Change Is 'Greatest Challenge Humans Have Ever Faced,' Author Says." Interview by Dave Davies. *Fresh Air*. Aired April 16, on NPR. https://www.npr.org/2019/04/16/713829853/climate-change-is-greatest-challenge-humans-have-ever-faced-author-says.

Metzl, Jamie. 2019. "Human Gene Editing Is Too Transformative to Be Guided by the Few." *Financial Times*, March 27, 2019. https://www.ft.com/content/6ebc7f3e-4ff5-11e9-8f44-fe4a86c48b33.

Milken Institute Staff. 2021. "Strive Masiyiwa: 'How Can I Say Science Has Delivered a Miracle to People Who Are Dying?'" June 25, 2021. https://milkeninstitute.org/video/strive-masiyiwa-science-delivered-miracle /.

Mills, Peter. 2019. "Three Venues for Discussing Human Gene Editing." *Issues Sci Technol* 35 (3). https://issues.org/three-venues-for-discussing-human-gene-editing/.

Mills, Peter. 2020. "Heritable Human Genome Editing: The National Academies and Royal Society Report." *Nuffield Council on Bioethics* (blog). September 28, 2020. https://www.nuffieldbioethics.org/blog/heritable-human-genome-editing-the-national-academies-royal-society-report.

Mullin, Emily. 2017. "The Fertility Doctor Trying to Commercialize Three-Parent Babies." *MIT Technology Review*, June 13, 2017. https://www.technologyreview.com/s/608033/the-fertility-doctor-trying-to-commercialize-three-parent-babies/?set=608091.

Mullin, Emily. 2019. "There Still Aren't Any Rules Preventing Scientists from Making Gene-Edited Babies." *OneZero*, November 25, 2019. https://onezero.medium.com/a-year-after-the-crispr-babies-debacle-a-rift-is-growing-among-the-worlds-scientists-48c8ded092bb.

National Academies of Sciences, Engineering, and Medicine (NASEM). 2017. *Human Genome Editing: Science, Ethics, and Governance*. Washington, DC: National Academies Press. https://doi.org/10.17226/24623.

National Academies of Sciences, Engineering, and Medicine (NASEM). 2019. *Human Genome Editing Initiative*. http://www.nationalacademies.org/gene-editing/index.htm.

National Academies of Sciences, Engineering, and Medicine (NASEM). 2019. "News Release: New International Commission Launched on Clinical Use of Heritable Human Genome Editing." https://www.nationalacademies.org/news/2019/05/new-international-commission-launched-on-clinical-use-of-heritable-human-genome-editing.

National Academy of Medicine, National Academy of Sciences, and the Royal Society. 2020. *Heritable Human Genome Editing*. Washington, DC: National Academies Press. https://www.nap.edu/catalog/25665/heritable-human-genome-editing.

Nature. 2019. "The Scientific Events That Shaped the Decade." December 18, 2019. https://www.nature.com/articles/d41586-019-03857-x.

Neergaard, Lauran. 2018. "AP-NORC Poll: Edit Baby Genes for Health, Not Smarts." *Associated Press*, December 28, 2018. https://apnews.com/general-news-ef1161deac194f2ca1fd99457dc2cf15.

New Scientist. 2019. "New Scientist Ranks the Top 10 Discoveries of the Decade." December 18, 2019. https://www.newscientist.com/article/mg24432613-200-new-scientist-ranks-the-top-10-discoveries-of-the-decade/#ixzz71NrrTpJP.

Nicol, Diane, Rebecca Paxton, Simon Niemeyer, Nicole Curato, John Dryzek, Christopher Rudge, Sonya Pemberton, and Francesco Veri. 2022. "Genome Editing: Formulating an Australian Community Response." Occasional Paper No 12, Centre for Law and Genetics. https://www.utas.edu.au/__data/assets/pdf_file/0011/1634258/OP12-final-report.pdf.

Nuffield Council on Bioethics. 2018. *Genome Editing and Human Reproduction*. London: Nuffield Council. http://nuffieldbioethics.org/project/genome-editing-human-reproduction.

Pappas, Stephanie. 2019. "The 10 Biggest Science Stories of the Decade." *Live Science*, December 27, 2019. https://www.livescience.com/biggest-science-of-the-decade.html.

Parens, Erik, and Adrienne Asch. 1999. "The Disability Rights Critique of Prenatal Genetic Testing: Reflections and Recommendations." *Hastings Center Rep* 29 (5): S1–22. https://doi.org/10.2307/3527746.

Piracés, Enrique. 2018. "Let's Avoid an Artificial Intelligentsia: Inclusion, Artificial Intelligence, and Human Rights." *Points Blog*, October 12, 2018. https:// medium .com/datasociety-points/lets-avoid-an-artificial-intelligentsia-inclusion-artificial -intelligence-and-human-rights-3905d708e7ed.

Ranisch, Robert. 2019. "Germline Genome Editing versus Preimplantation Genetic Diagnosis: Is There a Case in Favour of Germline Interventions?" *Bioethics* 34 (1): 60–69. https://doi.org/10.1111/bioe.12635.

Regalado, Antonio. 2018. "Rogue Chinese CRISPR Scientist Cited US Report as His Green Light." *MIT Technology Review*, November 27, 2018. https://www.technologyreview.com /2018/11/27/1821/rogue-chinese-crispr-scientist-cited-us-report-as-his-green-light/.

Regalado, Antonio. 2019. "We Won't Use CRISPR to Make Super-smart Babies— But Only Because We Can't." *MIT Technology Review*, January 18, 2019. https:// www.technologyreview.com/s/612774/we-wont-use-crispr-to-make-super-smart -babiesbut-only-because-we-cant/.

Saey, Tina Hesman. 2019. "A Nobel Prize Winner Argues Banning CRISPR Babies Won't Work." *Science News*, April 2, 2019. https://www.sciencenews.org/article/nobel-prize -winner-david-baltimore-crispr-babies-ban.

Saha, Krishanu, J. Benjamin Hurlbut, Sheila Jasanoff, Aziza Ahmed, Anthony Appiah, Elizabeth Bartholet, Françoise Baylis, et al. 2018. "Building Capacity for a Global Genome Editing Observatory: Institutional Design." *Trends Biotechnol* 36: 741–43. https://doi.org/10.1016/j.tibtech.2018.04.009.

Savulescu, Julian, and Guy Kahane. 2009. "The Moral Obligation to Create Children with the Best Chance of the Best Life." *Bioethics* 23 (5): 274–90. https://doi.org /10.1111/j.1467-8519.2008.00687.x.

Shanks, Pete. 2019. "Moratorium on Germline Gains Momentum." *Biopolitical Times*, May 9, 2019. https://www.geneticsandsociety.org/biopolitical-times/moratorium -germline-gains-momentum.

Sheehy, Cody, director. 2022. *Make People Better*. Rhumbline Media.

Silver, Lee M. 1997. *Remaking Eden: Cloning and Beyond in a Brave New World*. New York: Avon.

Silver, Lee M. 2000. "Reprogenetics: How Do a Scientist's Own Ethical Deliberations Enter into the Process?" Paper presented at Humans and Genetic Engineering in the New Millennium: How Are We Going to Get "Genetics" Just in Time? Copenhagen: Danish Council of Ethics, 2000. https://repository.library.georgetown.edu /handle/10822/523362.

Stein, Rob, presenter. 2015. "Scientists Debate How Far to Go in Editing Human Genes." *All Things Considered*. Aired December 3, 2015, on NPR. https://www.npr.org /sections/health-shots/2015/12/03/458212497/scientists-debate-how-far-to-go -in-editing-human-genes.

Sufian, Sandy, and Rosemarie Garland-Thomson. 2021. "The Dark Side of CRISPR." *Scientific American*, February 16, 2021. https://www.scientificamerican.com/article/the -dark-side-of-crispr/.

Sykora, Peter, and Arthur Caplan. 2017. "The Council of Europe Should Not Reaffirm the Ban on Germline Genome Editing in Humans." *EMBO Rep* 18: 1871–72. https://doi .org/10.15252/embr.201745246.

Travis, John. 2015. "Inside the Summit on Human Gene Editing: A Reporter's Notebook." *Science*, December 4, 2015. https://www.sciencemag.org/news/2015/12/inside-summit -human-gene-editing-reporter-s-notebook.

Turley, Patrick, Michelle N. Meyer, Nancy Wang, David Cesarini, Evelynn Hammonds, Alicia R. Martin, Benjamin M. Neale, et al. 2021. "Problems with Using Polygenic Scores to Select Embryos." *N Engl J Med* 385: 78–86. https://doi.org/10.1056/NEJMsr2105065.

Turnwald, Bradley P., J. Parker Goyer, Danielle Z. Boles, Amy Silder, Scott L. Delp, and Alia J. Crum. 2019. "Learning One's Genetic Risk Changes Physiology Independent of Actual Genetic Risk." *Nat Hum Behav* 3: 48–56. https://www.nature .com/articles/s41562-018-0483-4?fbclid=IwAR3uTtksOn6iyMRf5awv3lG8kbcoX9i -ZoaGylWlITBUxEPtNsr5dobsolE.

Viotti, Manuel, Andrea R. Victor, Darren K. Griffin, Jason S. Groob, Alan J. Brake, Christo G. Zouves, and Frank L. Barnes. 2019. "Estimating Demand for Germline Genome Editing: An *In Vitro* Fertilization Clinic Perspective." *CRISPR J* 2 (5): 304–15. http://doi.org/10.1089/crispr.2019.0044.

Vitez, Michael. 2015. "Thomas Gray Lived Six Days, but His Life Has Lasting Impact." *Inquirer*, March 29, 2015. https://www.philly.com/philly/health/20150329_Thomas _Gray_lived_six_days__but_his_life_has_lasting_impact.html.

Weaver, Matthew. 2018. "Private IVF Clinics Urged to Stop Charging for Expensive Add-Ons." *Guardian*, November 12, 2018. https://www.theguardian.com/society/2018 /nov/12/private-ivf-clinics-urged-to-stop-charging-for-expensive-add-ons.

WHO Expert Advisory Committee on Developing Global Standards for Governance and Oversight of Human Genome Editing. 2021a. *Human Genome Editing: Recommendations*. Geneva: World Health Organization.

WHO Expert Advisory Committee on Developing Global Standards for Governance and Oversight of Human Genome Editing. 2021b. *Human Genome Editing: A Framework for Governance*. Geneva: World Health Organization.

Williams, David R., and Lisa A. Cooper. 2020. "COVID-19 and Health Equity—A New Kind of 'Herd Immunity.'" *JAMA* 323 (24): 2478–80. https://doi.org/10.1001/jama .2020.8051.

World Health Organization (WHO). 2019. *Statement on Governance and Oversight of Human Genome Editing*. Geneva: World Health Organization. https://www.who.int /news/item/26-07-2019-statement-on-governance-and-oversight-of-human-genome -editing.

Yong, Ed. 2015. "What Can You Actually Do with Your Fancy Gene-Editing Technology?" *Atlantic*, December 2, 2015. https://www.theatlantic.com/science/archive/2015/12 /what-can-you-actually-do-with-your-fancy-gene-editing-technology/418377/.

3

Heritable Genome Editing and International Human Rights

Kevin Doxzen and
Jodi Halpern

On November 25, 2018, news broke of a scientist in China crossing a threshold that had never been traversed in human history (Regalado 2018). He Jiankui announced that he had genetically altered human embryos, leading to the birth of twin girls. Despite the absence of definitive evidence of the procedure, the news alone prompted widespread condemnation and calls by leading scientists for a moratorium on any further heritable genome editing (HGE) for reproductive purposes (Lander et al. 2019). In an effort to accelerate policy to keep pace with rapidly advancing genome editing technologies, national and international bodies called for the creation of clear and comprehensive guidelines that would articulate under what circumstances and within what boundaries HGE would be permissible (National Academies of Sciences, Engineering, and Medicine [NASEM] 2017; National Academy of Medicine, National Academy of Sciences, and the Royal Society 2020; Nuffield Council on Bioethics 2018a). While some authoritative bodies are focusing on HGE's safety and efficacy, we suggest an approach that begins with identifying serious concerns about social exclusion and social justice. These concerns, we argue, cannot be resolved by aiming to maximize positive over negative health outcomes. Rather, concerns raised by HGE reflect people's individual rights, rights that exist independently of population-level health outcomes. To prioritize human rights when deciding permissible uses of HGE, we recommend a tool that considers the possibility that genome-editing technology in some forms, not necessarily HGE, can improve society's collective health but focuses first on existing risks to individuals' core rights. We argue that the Human Rights Impact Assessment (HRIA) is suited to perform this necessary balancing act, underpinning an effective regulatory framework to guide the consideration of future uses of HGE (Gostin and Mann 1994).

Before developing an effective regulatory framework with an emphasis on human rights, we must identify the distinct features that trigger societal concerns over HGE. The rapid and widespread emergence of discussions surrounding the CRISPR babies case confirms that scientists' engineering of heritable changes to human beings touches on something core to the human experience. This kind of genome editing goes beyond conventional concerns about the safety and effectiveness of the technology. In contrast to HGE, applications of nonheritable genome editing tend to generate a more subdued response from both the public and the scientific or medical communities. For example, when undertaken in FDA-approved clinical trials, editing the *CCR5* gene in adults with the goal of treating HIV infection was heralded as an important medical step forward (US National Library of Medicine 2018). Though scientists involved in the CRISPR babies experiment targeted the same *CCR5* gene, aiming to prevent HIV infection, many saw this milestone as a step backward. This view does not indicate that all theoretical nonheritable genome editing applications are immune from societal concern. For example, genome editing for restoring hearing in deaf communities can be understood as a cultural threat and a disruptive reframing of deafness rather than a benefit (Scully 2008). Yet beyond these cases of polarizing nonheritable genome editing applications, HGE appears to encounter a volatile terrain of universal scrutiny and intense objection. To establish meaningful regulations, we must first understand the reasons behind these objections.

The Phantom Line between Treatment and Enhancement

One theme running through HGE regulatory recommendations, including the 2017 NASEM report on genome editing and the call for a moratorium on HGE published in *Nature* (Lander et al. 2019), suggests that treatment versus enhancement is a core societal concern. Polls have indicated public unease in using HGE to enhance a person rather than to address a severe unmet medical need. The NASEM and the moratorium authors refer to this distinction as a clear dividing line—what we refer to as a *permissibility pillar*—between what is allowed and what is off-limits (Funk and Hefferon 2018). However, a robust regulatory framework cannot stand upon such a distinction. The line between a treatment and an enhancement is blurry at best.

First, the concepts of *treatment* and *enhancement* are context dependent. A treatment in one cultural or geographic setting, or at one period in history, may be an enhancement in another. This shift is particularly true as new technologies like novel medications and advanced procedures become more mainstream. For example, society's use of cholesterol-lowering medications to prevent heart disease has increased over time. This change is driven by the medical community's updated metrics on what is considered healthy cholesterol levels, highlighting how society's health norms and standard of care evolve as new technologies, like statins and other cholesterol medications, are introduced. Because treatments may be defined as alterations that "restore [a person] to a normal state of human health and fitness" (President's Council on Bioethics 2003) and enhancements go beyond what is "normal," a constantly changing public health landscape can shift what was once an enhancement into the realm of normal medical care. Would germline editing of the *PCSK9* gene to lower cholesterol to permanently reduce cardiac risk show a similar trajectory from enhancement to treatment if such a procedure became the standard of care?

Second, using genetics to *prevent* disease does not fall clearly into either treatment or enhancement but instead further distorts the line between the two. Multiple reports have categorized this use of HGE as a therapeutic application, yet such preventive measures may lead to human enhancement in multiple ways. Juengst and colleagues (2018) provide an example of altering the genome to increase expression of the Klotho protein to prevent degenerative neurological conditions. Increased production of Klotho has also been shown to enhance cognition in mice, an example of an "incidental enhancement." Preventive genome editing to increase Klotho expression has the primary goal of minimizing the risk of disease and the secondary byproduct of enhancement. The problem of defining prevention arises when minimizing disease risk and also appears with targeted treatments, especially as the concept of a disease changes. For example, the idea of classifying aging as a disease is gaining support (Adam 2019). Slowing the process of aging would currently be considered an enhancement, but advancements in biological understanding and genome-editing technology may change the definition of aging. Targeted genome editing of telomerase reverse transcriptase genes to minimize aging may be considered a germline treatment for a disease in the future (Tomás-Loba 2008). There are reasons such preventive or even targeted therapeutic germline edits are ethically problematic, which we turn to next, but those reasons cannot be pinpointed by drawing an enduring distinction between enhancement and treatment. Instead, an ethical regulatory framework for HGE must be built on cross-cultural and enduring values.

Social Justice and the Threat of Eugenics

We need a much more robust scrutiny of what makes for potentially valuable preventive alterations, such as editing the *PCSK9* gene to lower cholesterol or the *CCR5* gene for HIV prevention, ethically unacceptable within germlines rather than in somatic cells. We see such a determination stemming from two concerns, one about social inclusion and the threat of eugenics, and the other about social justice. Both concerns were raised in an influential set of guidelines proposed by another national academic body, the Nuffield Council. The Nuffield Council on Bioethics report on genome editing and human reproduction stressed that HGE may be ethical if an alteration upholds "principles of social justice and solidarity, i.e., it should not be expected to increase disadvantage, discrimination, or division in society" (Nuffield Council 2018b). Crucially, these are rights-based concerns. They depend on the fundamental right of each person to be treated with equal regard and respect, which includes that each person has a fair chance at receiving a necessary and serious health benefit and a right to be protected from discrimination and social exclusion. We agree with the Nuffield Council that these rights-based concerns are essential, requiring that any regulatory framework for HGE centers around human rights. In arguing for such a framework, let us describe how HGE poses threats to the concerns of social exclusion and social justice.

We must confront the issue of social exclusion when considering how HGE may lead to the increased stigmatization of people with disabilities and even a potential slide into eugenics. HGE involves selecting out certain traits with the assumption that these traits make lives lesser lives. While this may seem like an understandably motivated selection against degenerative diseases like Huntington's, it could proceed into selection against people with chronic disabilities who view their own lives as quite worth living but whom society devalues (Sufian and Garland-Thomson 2021).

The paradigm case that has been raised is that of deafness. Although the medical community has viewed deafness as a disability worth preventing through cochlear implants, some deaf activists have seen the implants as an unwarranted intrusion that threatens the existence of their valued culture (including sign language and tightly knit social communities) (Aronson 2000). Many other deaf people and families approve of cochlear implants and appreciate this intervention. Perhaps the possibility of a somatic, post birth genome edit, which individual families could independently decide on, would be greeted with similar mixed responses. On the other hand, an HGE intervention is likely to create a different level of pressure to eliminate all deafness, a situation that resonates with the history of eugenics. Within an environment that allows the use of HGE for deafness, children born deaf could be even more isolated and stigmatized, and deaf culture extinguished. As an extension of this example, many other health differences could eventually become selected against through HGE, ranging from visual impairment to cognitive differences, leading to new targets of eugenics.

The fact that HGE raises the threat of eugenics might be considered a full stop on proceeding with HGE, but that has not been the initial instinct of society. First, selection based on genetic traits is already accepted in the practice of preimplantation genetic diagnosis (PGD). PGD involves the process of genetically screening multiple embryos to decide which ones are fit for implantation based on their genetic makeup. Some ethicists argue PGD is more offensive than HGE because it selects among potential lives (rather than among traits) to allow only embryos without the specified trait to develop. Despite this embryonic selection process, public protests about PGD are infrequent, in part because PGD is not likely to scale to the level of overwhelming societal effects any time soon (*Gènéthique* 2015). PGD requires in vitro fertilization (IVF), a procedure with relatively low "success" rates, high price tags, and limited accessibility to much of society. Notably, though, HGE also requires IVF. Given that in 2012, only 1.5% of US births were conceived via IVF, it is unlikely that IVF rates will rapidly expand and incorporate HGE to a point of having a substantial species-level impact (CDC et al. 2014). Nevertheless, the costs for innovative technologies often decrease, while accessibility and success rates may improve over time. In either case, if those with the resources to access the technology increasingly used HGE to screen against and genetically edit out disabilities, any such use is likely to further exacerbate the social stigmatization and exclusion that marginalized communities experience.

Inequality exacerbated by unequal access to HGE and use of genome editing as an alternative reproductive technology led to our second major concern, one of broader social justice. Specifically, we define this issue of social justice as an unfair aggregation of benefits, further described below. The sophisticated technical requirements of HGE, along with the necessity of using IVF for implementation, assure that for the near future, access to the technology will be very costly. HGE would likely be another privilege of the wealthy, providing more health and other advantages that will remain out of reach for most of society.

There are important ethical arguments against aggregating too many forms of advantage in some people. Such concerns apply even more strongly to aggregating advantages in families across generations. Michael Walzer (1983) argued in *Spheres of Justice* that while a capitalist society could tolerate groups having dominance in some aspects of life—some in wealth, some in health, and some in education or civic leadership—it is patently unjust for some groups to have dominance across spheres. We know from research on the structural determinants of health that some groups already have entirely too much control over multiple spheres. For example, economic advantages can lead to educational and health advantages. Surely, any financial barrier to HGE would compound this aggregation of benefits significantly.

Correspondingly, societal use of HGE could mean that individuals already living with disadvantages in one sphere could experience the aggregation of disadvantages across multiple spheres. For example, curative technologies to treat diseases can contribute to the degradation of disability cultures and the disrespect of persons (Sufian and Garland-Thomson 2021). Even within a single community, members can experience the aggregation of disadvantages. For the deaf community, the use of cochlear implants to "cure" deafness means that the fraction of individuals who are unable to access curative technologies or voluntarily choose to remain deaf may accumulate disadvantages as support provided by deaf culture diminishes (Sparrow 2005).

Thus, we come to an important recognition regarding the ethics of HGE. In our view, the more serious ethical concerns arise because of existing intergenerational injustice in the distribution of health and other benefits in our society. HGE, like many other expensive, innovative technologies, is poised to amplify such injustice. Such a conclusion does not mean that we must reject market economies. In fact, capitalism as we know it arose as an alternative to the dynastic, heritable control of wealth. Nobel Prize winner Amartya Sen (1979) has addressed the question of what comprises the basis of fairness in a free and just capitalist society. Sen and G.A. Cohen (1989), perhaps the greatest thinker extending Sen's work, agree that fairness depends on equality of access to the advantages needed to pursue a good life. Rejecting a heritable familial aggregation of benefits can be seen as central to democratic values.

Recognizing that HGE may further widen the gap between the haves and have-nots and stigmatize oppressed and vulnerable communities, there remain scenarios in which genetic alteration of the human germline might be ethically acceptable. An ideal regulatory framework would consider the future life of the edited embryo and the broader circumstances of society. This holistic view is achieved under the purview of human rights.

A Human Rights Foundation for Ethical Frameworks

How do we move forward? Notably, a consistent thread appears among the various scientists calling for moratoria, self-policing, or other delays in HGE, as well as in the NASEM (2017) and Nuffield (2018b) reports. These statements call for robust public debate that includes a broad spectrum of voices. This proposed exercise in popular governance of technological innovation has merit in its inclusion of those impacted in the decision-making process, but simply creating the opportunity to debate does not ensure the application of social justice, inclusion, and solidarity. Rather, broad debate runs the risk of simply replicating implicit biases, with majority or more empowered groups dominating others. For example, able-bodied voters in Oregon inadvertently ranked disability-related outcomes so low that disabled people were excluded from treatments (Bickenbach 2016). To prevent such dominance, it is essential to have transparent frameworks for ethical deliberation and decision-making—that is, explicit criteria and assessment tools. Further, any attempt at including all possible and emerging voices on equal ground without an assessment tool to guide the conversation could lead to an ongoing debate without a tangible outcome.

Despite these caveats, public engagement is imperative, and several groups, like the United Kingdom's National Co-ordinating Centre for Public Engagement (NCCPE) and the Royal Society, have already begun approaching defined stakeholders and broad audiences (NCCPE 2018; van Mill, Hopkins, and Kinsella 2017). Beyond receiving stakeholder input, translation into just and actionable policy requires robust and transparent ethical frameworks. These frameworks are necessary to set boundaries for societal debate. Their exact use must also be made public so they are applied consistently and fairly. Thus, we would disagree with the *Nature* (NASEM 2017) commentary, which places hope in a five-year waiting period during which countries are expected to establish their own regulatory frameworks. Importantly, the commentary avoids suggesting any standards for what would count as ethical progress. The authors state that "the governance model we present would intentionally leave room for nations to take differing approaches and reach different conclusions, informed by their history, culture, values and political systems. Still, the common principle would be all nations agreeing to proceed deliberately and with due respect to the opinions of humankind" (168).

We place less faith in the opinions of humankind and the politically expedient values of particular countries at particular times and more faith in international standards for human rights as a guide for approaching how to use CRISPR for HGE applications. Abundant research in psychology, political science, and behavioral economics shows how public opinion can be highly unstable and easily manipulated (Zaval and Cornwell 2016). But human rights stand above the political fray and set standards that protect every member of society. Notably, technologies like HGE can readily cross borders, and wealthier people can travel to obtain them. Thus, we agree with Zeid Ra'ad Al Hussein (2019), former UN High Commissioner of Human Rights, that the ethical implementation of border-crossing technologies (artificial intelligence in his editorial) should not be ruled by local cultures or prevailing attitudes but guided by a commitment to improving universal human rights. To this end, one framework for assessing the ethics of using CRISPR is to adapt the Human Rights Impact Assessment (HRIA) for addressing specific societal scenarios regarding HGE (Gostin and Mann 1994).

The HRIA provides a public health approach for assessing whether the benefits of a given HGE application, considering the societal context, outweighs any infringement on human rights. This tool helps to identify when a specific use of HGE may disproportionately impact vulnerable communities. The HRIA is a compilation of seven questions, the specific details of which are beyond the scope of this essay but discussed elsewhere (see Halpern et al. 2019). The questions incorporate considerations of intrusiveness, efficacy, and targeting appropriate populations to avoid exploitation or exclusion. Practitioners of the HRIA have demonstrated its feasibility in areas that share ethical and public health features with the challenges of HGE, specifically the health rights of women (Bakker 2009). Other researchers have argued for a human rights-based approach to regulating HGE, but they begin with "the human right to science," which focuses on the specific application of HGE rather than the broader impact on society (Boggio 2019).

There are some notable implications of choosing the HRIA as a core framework. First, it embodies a deontological, or rights-based approach to ethics, which is distinct from and can conflict with a utilitarian approach. For example, under a utilitarian regulatory framework, an HGE application would be permissible if it maximizes a collective health outcome at a population level, either through preventive genome editing like reducing the risk of neurological diseases or through targeted therapeutic approaches like editing a gene to lower cholesterol. A deontological regulatory framework goes one step further and says that an HGE application would be impermissible if it maximizes a collective health outcome but also violates the rights of distinct groups of people, such as people with disabilities, or if it created an irreversible unfair burden on the economically least well off. Second, its focus on human rights also aligns with the concerns of the Nuffield Council and helps establish constraints on applications of HGE across diverse cultures and societal norms.

Third, a key aspect of using an HRIA to evaluate the permissibility of HGE is the case-by-case nature of the approach. Under a utilitarian approach, when HGE's permissibility depends on maximizing benefit over risk, incurable conditions like Tay-Sachs may be edited regardless of unfair access based on the cost of such a procedure. In contrast, implementation of an HRIA would consider other available options, like improving access to genetic screening for Tay-Sachs, that might meet the needs of all and avoid opening the door to germline genome editing. Beyond health and wellness, the HRIA considers factors impacting population-level justice and equality, factors that are usually outside the scope of utilitarian approaches.

We understand that the *Nature* authors were rightfully observant of the problem that any substantive ethical framework, including an HRIA, may not be accepted by some countries (Lander et al 2019). Our response to this concern is that such explicit frameworks are nevertheless essential to make genuine ethical progress. Recent recommendations by prominent scientific bodies have highlighted the critical role human rights should play in the development of future governance and oversight policies. The report on HGE by the National Academy of Medicine, National Academy of Sciences, and the Royal Society (2020) states, "the need to develop governance approaches to encompass HHGE [Heritable Human Genome Editing] provides a potential opportunity to use and develop the content of internationally recognized human rights to influence future laws, policies, and regulatory responses around HHGE." The governance framework on human genome editing developed by the World Health Organization (WHO) expert advisory suggests that any new international law around HGE could leverage existing declarations, treaties, and conventions addressing human rights, such as the

Council of Europe Convention on Human Rights and Biomedicine in Oviedo, Spain (Council of Europe 1999; WHO Expert Advisory Committee on Developing Global Standards for Governance and Oversight of Human Genome Editing 2021).

We urge new bodies, such as the Global Observatory and the Association for Responsible Research and Innovation in Genome Editing (ARRIGE) Initiative (Jasanoff and Hurlbut 2018; Montoliu et al. 2018) to commit to a human rights foundation for their deliberations. Rather than be distracted by the novel issues raised by HGE, it is crucial to consider how actual human lives and rights will be affected. Given the universal problems of social exclusion and unfair access to advantage, the implementation of HGE is sure to create human-rights challenges that must be addressed for any agenda in ethics to make a real improvement in human lives.

Society-changing technologies, by their very nature, instill both optimism and trepidation. Development of CRISPR and other genome editing technologies may lead to treatments of previously incurable genetic diseases. Yet even with principled intentions, deployment of genome-editing tools raises serious concerns, especially in the realm of HGE. As researchers, we believe we have an indispensable duty to guide the implementation of new technologies in ways that respect fundamental human rights. We fervently implore national and international bodies to integrate ethical and human rights considerations into their own evolving regulatory frameworks.

REFERENCES

Adam, David. 2019. "What If Aging Weren't Inevitable but a Curable Disease?" *MIT Technology Review*, August 19, 2019. https://www.technologyreview.com/s/614080/what-if-aging-werent-inevitable-but-a-curable-disease/.

Al Hussein, Zeid Ra'ad. 2019. "Here's What Will Decide Whether Technology Becomes a Force for Good, or Evil." *Washington Post*, April 9, 2019. https://www.washingtonpost.com/opinions/technology-can-be-put-to-good-use--or-hasten-the-demise-of-the-human-race/2019/04/09/c7af4b2e-56e1-11e9-8ef3-fbd41a2ce4d5_story.html.

Aronson, Josh, director. 2000. *Sound and Fury*. Public Policy Productions. 80 min.

Bakker, Saskia, Marieke Van Den Berg, Deniz Düzenli, and Marike Radstaake. 2009. "Human Rights Impact Assessment in Practice: The Case of the Health Rights of Women Assessment Instrument (HeRWAI)." *J Hum Rights Pract* 1 (3): 436–58.

Bickenbach, Jerome. 2016. "Disability and Health Care Rationing." In *Stanford Encyclopedia of Philosophy*. https://plato.stanford.edu/entries/disability-care-rationing/.

Boggio, Andrea, Bartha M. Knoppers, Jessica Almqvist, and Cesare P. R. Romano. 2019. "The Human Right to Science and the Regulation of Human Germline Engineering." *CRISPR J* 2 (3): 134–42. https://doi.org/10.1089/crispr.2018.0053.

(CDC) Centers for Disease Control and Prevention, American Society for Reproductive Medicine, and Society for Assisted Reproductive Technology. 2014. *Assisted Reproductive Technology National Summary Report*. Atlanta: US Department of Health and Human Services.

Cohen, Gerald Allan. 1989. "On the Currency of Egalitarian Justice." *Ethics* 99 (4): 906–44.

Council of Europe. 1999. "Oviedo Convention and Its Protocols." https://www.coe.int/en /web/bioethics/oviedo-convention.

Funk, Cary, and Meg Hefferon. 2018. "Public Views of Gene Editing for Babies Depend on How It Would Be Used." Pew Research Center. https://www.pewresearch.org /science/2018/07/26/public-views-of-gene-editing-for-babies-depend-on-how-it -would-be-used/.

Gènéthique. 2015. "Switzerland: Disabled Persons' Organisations Adopt a United Front against PGD (Pre-Implantation Genetic Diagnosis)." June 9, 2015. https://www .genethique.org/switzerland-disabled-persons-organisations-adopt-a-united-front -against-pgd-pre-implantation-genetic-diagnosis/?lang=en.

Gostin, Lawrence, and Jonathan M. Mann. 1994. "Towards the Development of a Human Rights Impact Assessment for the Formulation and Evaluation of Public Health Policies." *Health Hum Rights* 1 (1): 58–80.

Halpern, Jodi, Sharon E. O'Hara, Kevin W. Doxzen, Lea B. Witkowsky, and Aleksa L. Owen. 2019. "Societal and Ethical Impacts of Germline Genome Editing: How Can We Secure Human Rights?" *CRISPR J* 2 (5): 293–98. https://doi.org/10.1089 /crispr.2019.0042.

Jasanoff, Sheila, and J. Benjamin Hurlbut. 2018. "A Global Observatory for Gene Editing." *Nature* 555: 435–37.

Juengst, Eric T., Gail E. Henderson, Rebecca L. Walker, John M. Conley, Douglas MacKay, Karen M. Meagher, Katherine Saylor, et al. 2018. "Is Enhancement the Price of Prevention in Human Gene Editing?" *CRISPR J* 1 (6) 351–54. https:/doi.org/0.1089/crispr .2018.0040.

Lander, Eric S., Françoise Baylis, Feng Zhang, Emmanuelle Charpentier, Paul Berg, Catherine Bourgain, Bärbel Friedrich, et al. 2019. "Adopt a Moratorium on Heritable Genome Editing." *Nature* 567: 165–68.

Montoliu, Lluis, Jennifer Merchant, François Hirsch, Marion Abecassis, Pierre Jouannet, Bernard Baertschi, Cyril Sarrauste de Menthièr, and Hervé Chneiweiss. 2018. "ARRIGE Arrives: Toward the Responsible Use of Genome Editing." *CRISPR J* 1 (2): 128–30.

National Academies of Sciences, Engineering, and Medicine (NASEM). 2017. *Human Genome Editing: Science, Ethics, and Governance*. Washington, DC: National Academies Press. https://doi.org/10.17226/24623.

National Academy of Medicine, National Academy of Sciences, and the Royal Society. 2020. *Heritable Human Genome Editing*. Washington, DC: National Academies Press. https://doi.org/10.17226/25665.

National Co-ordinating Centre for Public Engagement (NCCPE). 2018. "Genome Editing Public Engagement Synergy." https://www.publicengagement.ac.uk/nccpe-projects -and-services/nccpe-projects/genome-editing-public-engagement-synergy.

Nuffield Council on Bioethics. 2018a. *Genome Editing and Human Reproduction: Social and Ethical Issues*. London: Nuffield Council.

Nuffield Council on Bioethics. 2018b. *Summary of Report: Genome Editing and Human Reproduction: Social and Ethical Issues*. London: Nuffield Council.

President's Council on Bioethics. 2003. *Beyond Therapy: Biotechnology and the Pursuit of Happiness*. Washington, DC: President's Council. https://bioethicsarchive.georgetown.edu/pcbe/reports/beyondtherapy/fulldoc.html.

Regalado, Antonio. 2018. "Chinese Scientists Are Creating CRISPR Babies." *MIT Technology Review*, November 25, 2018. https://www.technologyreview.com/s/612458/exclusive-chinese-scientists-are-creating-crispr-babies.

Scully, Jackie Leach. 2008. *Disability Bioethics: Moral Bodies, Moral Difference*. Lantham, MD: Rowman & Littlefield.

Sen, Amartya. 1979. "Equality of What?" Stanford University: Tanner Lectures on Human Values. https://www.ophi.org.uk/wp-content/uploads/Sen-1979_Equality-of-What.pdf.

Sparrow, Robert. 2005. "Defending Deaf Culture: The Case of Cochlear Implants." *J Polit Philos* 13 (2): 135–52.

Sufian, Sandy, and Rosemarie Garland-Thomson. 2021. "The Dark Side of CRISPR." *Scientific American*, February 16, 2021. https://www.scientificamerican.com/article/the-dark-side-of-crispr/.

Tomás-Loba, Antonia, Ignacio Flores, Pablo J. Fernández-Marcos, María L. Cayuela, Antonio Maraver, Agueda Tejera, Consuelo Borrás, et al. 2008 "Telomerase Reverse Transcriptase Delays Aging in Cancer-Resistant Mice." *Cell* 135 (4): 609–22.

US National Library of Medicine. 2018. "CCR5-Modified CD4 T Cells for HIV Infection (TRAILBLAZER)." ClinicalTrials.gov. National Institutes of Health (NIH). https://clinicaltrials.gov/ct2/show/NCT03666871.

van Mill, Anita, Henrietta Hopkins, and Suzanna Kinsella. 2017. *Potential Uses for Genetic Technologies: Dialogue and Engagement Research Conducted on Behalf of the Royal Society*. London: Royal Society. https://royalsociety.org/~/media/policy/projects/gene-tech/genetic-technologies-public-dialogue-hvm-full-report.pdf

Walzer, Michael. 1983. *Spheres of Justice: A Defense of Pluralism and Equality*. Oxford: Robertson.

WHO Expert Advisory Committee on Developing Global Standards for Governance and Oversight of Human Genome Editing. 2021. *Human Genome Editing: A Framework for Governance*. Geneva: World Health Organization.

Zaval Lisa, and James F. M. Cornwell. 2016. "Cognitive Biases, Non-Rational Judgments, and Public Perceptions of Climate Change." In *Oxford Research Encyclopedia of Climate Science*. November 2016. https://doi.org/10.1093/acrefore/9780190228620.013.304.

4

Democratizing CRISPR

Opening the Door or Pandora's Box?

Ellen D. Jorgensen

Full disclosure: I am not an ethicist. I am not one of the discoverers of CRISPR technology. I have never used CRISPR in my job as a molecular biologist. I did, however, codesign a hands-on class showing people how to use it. For years, I facilitated teaching CRISPR to anyone who walked in from the street and wanted to learn.

How could you be afraid of CRISPR if you use it in a neighborhood setting, side by side with your teenage daughter? This was the logic behind our 2015 decision to teach classes for the public in a community lab setting in Brooklyn. If you have not yet encountered the phenomenon known as a *community lab*, they are fully equipped biotechnology laboratories with a mission to serve the public by engaging them with hands-on experiments. Our decision to teach CRISPR techniques to anyone who wanted to learn was an act of defiance against CRISPR's predominant media portrayal as a dangerous technology that might need to be suppressed or banished (Maxmen 2015; Wadhwa 2015).

The 2012 demonstration of CRISPR's gene-editing potential by Emmanuelle Charpentier and Jennifer Doudna (Jinek et al. 2012) was a major event in scientific circles, making it hard for those who have chosen science as a career to understand why a foundational new technology can generate so little public interest outside of the fear that genetic engineering of humans might become commonplace. Scientists are part of an insular club, and although we are trying hard to be better communicators, our educational background and career focus often isolate us from mainstream life. Progress is measured in years or even decades, and the implications of our work may not be readily apparent even to ourselves. The Spanish scientist Francisco Mojica is widely acknowledged to be the first person to study the ubiquitous

CRISPR repeats in microbes and propose their function as a bacterial immune system, yet it took years before his findings were considered relevant outside of a niche audience of microbiologists (Lander 2015). It is often impossible for even experts in the field to know what will have a major impact and what will not, owing to the cumulative nature of scientific progress. The societal implications of discoveries are seldom clear at the beginning, and the public is normally not involved in shepherding new technologies into areas that align with their needs.

It makes perfect sense that earth-shaking scientific discoveries are largely ignored by most of the population. Unless they have an immediate and direct effect on people's day-to-day existence, most folks have enough hurdles to face without adding technology angst to the pile. I challenge you to walk up to someone on the street in an average neighborhood (not an academic campus or biotech hub) and ask them if they have heard of "CRISPR." Unless they are one of the few individuals who avidly follow breakthroughs in science, they likely will not know what you are talking about. In mainstream media, CRISPR news was inflammatory but fleeting. Those not faithfully paying attention might have missed the flurry of reports around the time of the CRISPR Asilomar meeting (Reardon 2015) and the announcement of China's edited babies (Normile 2018). GMO food labeling of "CRISPRed" mushrooms (Waltz 2016) and the first demonstration of gene drives in mosquitoes (Matthews 2018) might not even have made the news. If we genuinely want the public to help shepherd technology, then the first step is to get people to pay attention to it.

Though an important part of the mission of community lab spaces was to make the public aware of biotech advances, an equally important part was to prevent unnecessary fears by normalizing biotechnology via democratization and emphasizing its potential for good. Normally, the information flow for new science is a talking head in a white coat on television, surrounded by the alien environment of a professional lab space with its pristine benches, beakers, and mysterious machines. "We can't put it back into the box" is a typical phrase used to describe everything from AI to the demise of civil discourse in the political arena. The phrase signals a kind of fatalism, a feeling that a technology you do not understand and are not in control of is taking over aspects of your daily life without your consent. The typical response to this loss of control is suspicion and fear. The anti-GMO movement and the current antivaccine sentiment are symptoms of this response, and advances in biology such as the improved outcomes for cancer patients due to immunotherapies do little to counteract the bad rap biotech receives. When was the last time you saw a movie in which genetic engineering saved the day? Combating the dystopian image of DNA science was a driving force in the emergence of community labs. If the first part of the mission was to help the public become aware of CRISPR, the second was to engage them physically with the technology to normalize it.

What if there were a place embedded in your community where you could go to learn and experience biotechnology at its basic level? Community labs remove biotechnology from conventional settings, like universities and biotech companies, and bring it into more casual and friendly environments, like arts organizations and makerspaces. They democratize biotechnology through open classes and memberships, providing an inexpensive space where citizen scientists can pursue their own projects. Nothing is off the table if the proposed activity passes basic safety guidelines. The combination of kit-based protocols, the huge drop in cost and widespread availability of DNA synthesis, and the rise of the maker movement (when homemade versions of simple lab equipment became a reality) made the attainment of community labs easy. We made a deliberate choice to engage the public through hands-on experiments, because, in my experience, nonscientists successfully using technology can help de-escalate fears. Basic biotech protocols, now decades old, are kids' stuff. Most parents are unaware that in advanced high school biology classes, their children perform the classic bacterial transformation of a plasmid containing a green fluorescent protein gene into *E. coli* bacteria, creating an organism that glows green under UV light because it now contains a gene stolen from a jellyfish. Many of the Nobel prize-winning experiments of the past are now part of advanced biology curricula and educational kits sold to schools globally. Logically, CRISPR should follow this pattern, but it posed a dilemma. Given the history of community labs, especially our love-hate relationship with the press (save the world or destroy the world in what Drew Endy of Stanford has termed the *half-pipe of doom*) we tended to shy away from potential controversy for self-preservation.

In 2015, we began to get inquiries from the press. Were we using CRISPR? Were we editing ourselves? Could anyone do it? CRISPR was widely touted as "cheap and easy," and that was enough to fuel the fear that amateurs might try their hand at playing God. This fear was something we had faced in the past. Each sufficiently powerful discovery in biology seemed like an easy way to generate a news article by speculating on the damage that could be caused by so-called amateurs getting their hands on the technology. When the H5N1 flu variant was a concern, articles were published calling community labs a potential threat to manufacture it (Zimmer 2012). It was logical to expect the same kind of treatment in the press but focusing this time on CRISPR. There was a real possibility that teaching CRISPR to nonscientists could reopen the old debate around the value of community-lab spaces versus the perceived risk they posed. With that in mind, we began the internal debate around whether it would harm or help our mission to open this technology to the public.

What were the positive outcomes of making this technology accessible to everyone? CRISPR was bursting with possible uses unrelated to engineered babies. It was unfortunate that all the negative media hype had effectively obscured the reality that CRISPR was incredibly transformational to standard biology workflows. The technology enabled researchers to precisely delete or alter DNA sequences in a cheaper, faster way without changing the fundamental purpose and goals of the experiments. It could speed up the type of conventional genetic-engineering projects at the heart of the industry that had been continuing in academic and industrial labs since the 1980s to produce cures for diseases, healthier crops, and industrial enzymes in a renewable manner. New variants of CRISPR were opening the door to new diagnostic tests and therapies. Though we did not want our actions to feed into the stereotype of amateurs doing dangerous things, we strongly felt that the public should engage with these important advances (Wang and Doudna 2023). As with other genetic engineering methods, being able to interact with CRISPR in a community setting could help demystify it. A more general understanding of the technology at a hands-on level might allow people to see it as a tool with no inherent good or evil quality. The question was, should we avoid the controversy or lean in?

At the same time, people at the fringes of the citizen-science community began capitalizing on the media hype around CRISPR. One enterprising individual began selling DNA components of the CRISPR system online with claims that anyone could use it to cure disease. He posted (and was later forced to remove) controversial videos on YouTube where he attempted to edit the DNA in his skin cells by abrading his arm and applying CRISPR DNA (Lee 2017). He later received warning letters from the FDA. Most of our community was appalled by these "self-experiments." We strove to distance ourselves from this irresponsible and deceptive behavior. It was Tom Knight, acknowledged as the father of synthetic biology and one of our most respected board members, who convinced my community lab that this unfortunate development gave us an even greater reason to go forward. He pointed out that by being thoughtful about the classes we designed to teach about CRISPR, we could frame ourselves as the "responsible CRISPR lab." We could be true to our mission but not add to the misinformation and confusion around the technology. This idea became the framework for the outreach we did around CRISPR for the next several years.

Some people exhibit enthusiasm for acquiring technical skills but appear much less interested in learning about the broader implications of the technology they practice. In our classes, we always strove to take a broader view that included discussion of how society would be affected by biotechnology. We were uncomfortable with the idea of teaching CRISPR as a technique without context, so we developed a two-part curriculum in which an hourlong lecture was a prerequisite to the lab course. The lecture covered how CRISPR works but also comprehensively covered the history of its discovery and the social, moral, and ethical challenges posed by gene editing. Ours was one of the first organizations to offer a CRISPR class, and the very first to teach CRISPR techniques to anyone who wanted to learn them. To our surprise, a substantial proportion of the people enrolling in our class were conventionally trained scientists who could not find anyone else offering hands-on instruction in what they knew to be a new foundational technique. We were more than happy to share our knowledge with them.

So, what did the hands-on laboratory class cover? CRISPR was neither cheap nor easy to use by community lab standards. The fact that "cheap and easy" was an expensive and complicated process was never mentioned in news reports, yet even someone with a great deal of training in laboratory techniques would find using CRISPR difficult at first. We decided the class should tackle something basic but fun. We wanted to replicate a classic CRISPR demonstration published by George Church's lab at Harvard (DiCarlo et al. 2013). The protocol had several attractive features: it was a simple experiment in harmless bakers' yeast, and it was a gene knockout that produced a color change that could be seen easily. At the time, no CRISPR kits were commercially available; all the reagents and supplies for the work had to be assembled from various sources.

We boiled down the experiments in Church lab's publication to a simple one that could be completed within the framework of four 3-hour classes. We bought CRISPR-containing plasmids (circular pieces of DNA that would deliver the editing system) from a nonprofit repository. We purchased yeast strains and media components. To ensure that the students thoroughly understood the process, we put them through the steps of target design and modification of guide RNA sequences to customize CRISPR for the knockout we were aiming for. They built and purified the materials they would use for the knockout. We later learned that the class included more actual hands-on work than similar classes that appeared afterward in academic institutions. Eventually, we developed a minicourse that we delivered on-site to high schools and science outreach events, always careful to embed the societal impact discussions within the activity.

So where do we go from here? CRISPR technology has gone through multiple iterations since we developed the community lab class for the public. It has become easier to use. Despite helpful advice from the Church lab, it was months before the PhD-level scientists within our community lab were able to tweak and perfect the class activity such that consistent success was achieved by most of the students. The "easy" technology was anything but easy in its initial incarnation. Currently, plasmids encoding CRISPR are not difficult to obtain and are even sold in kits marketed to the DIY community. This seems alarming but having taught hundreds of citizen scientists how to use the tools of molecular biology, I have observed that very few want to spend the time and effort it takes to successfully engineer organisms. There is still a sizable barrier to entry based on the amount of basic background knowledge necessary to successfully use these tools—everything from doing concentration calculations to using a pipette. This is not to say that nobody will succeed, but in my view, the danger is like that posed by putting older genetic engineering methodology into the hands of amateurs, a practice that has been going on in community labs for over ten years with no adverse events. We were careful to contextualize the CRISPR technology in our classes, pointing out that there could be off-target edits and describing how professional scientists were proceeding with caution in using CRISPR in a medical context. Given that the most controversial aspect of CRISPR is human embryo editing, the DIY community is a wonderful place to have ethical discussions around this use. I believe that eventually society will make peace with the idea of editing human embryos to eliminate certain specific genetic diseases in the germline, and I applaud the serious discussions going on in the science and ethics community around this possibility. However, it is unreasonable to fear that citizen scientists would somehow engineer babies (I have yet to see a community lab that includes an in vitro fertilization facility).

A more pertinent area of concern is that amateurs might attempt to engineer their own tissues by mimicking a series of irresponsible YouTube videos (since taken down) of citizen scientists attempting to apply the technology dermally. Although none of these efforts appears to have succeeded, the growing public distrust of the pharmaceutical and medical industries (including the slow and careful process by which new therapies are approved by the FDA) has fueled the idea of taking individual action to "cure" diseases with genetic roots. Following rigorous clinical testing in conventional institutional environments, somatic cell gene editing is already current practice in therapies for cancer and blood diseases but trying it at home is an exercise in futility, especially given the rapid pace of research. The greatest danger I see is that someone using a kit without the support of the expertise found in community labs will not fully understand the science and manage to hurt themselves in the process. Anyone hawking DIY CRISPR kits for human or animal use should be shut down, because it is highly unlikely that they will provide any tangible medical benefit to the user who might be desperate for a solution to their own or a loved one's disease. But let us not equate the responsible use of CRISPR in community labs with self-experimentation or embryo engineering, although these labs are an excellent place to continue the discussion around it (Jorgensen and Grushkin 2011; Scheifele and Burkett 2016).

Any discussion of CRISPR is not complete without mentioning self-perpetuating incarnations of CRISPR, such as gene drives. These have the capability of rapidly spreading through a population, do not belong in community labs, and perhaps do not belong on the planet at all, because they can disrupt entire ecosystems (Kahn 2016). There has been remarkable progress in modifying gene drives to make them more efficient and more controllable, but field testing in isolated areas, like islands, is needed to evaluate them from a risk-reward perspective. The prospect of eliminating mosquito-borne diseases such as malaria is attractive, but thankfully, the biomedical community and regulatory bodies are proceeding with extreme caution.

Engineering insects requires sophisticated techniques and microinjection equipment not available for amateur use. There is little danger that citizen scientists will create and release gene drives. If that event occurs, there have been significant advances within the professional community around how to stop a gene drive from going wrong. Despite suggested guidance drafted by committees at the National Institutes of Health, the United Nations, and other supervisory organizations, however, there are currently no laws in place limiting the creation of gene drives (Global Gene Editing Regulation Tracker 2023). Their release into the environment is governed by regulatory bodies such as the US Environmental Protection Agency, but the process of building a drive remains unregulated. Further discussion of gene drives is outside the scope of this essay, but this is an area where progress needs to be carefully monitored. Awareness of gene drives is the first step in educating the public and allowing them to participate in discussions from a position of understanding, not fear. Community labs can spearhead this effort by holding public conversations around this subject.

After witnessing hundreds of individuals discover the magic of gene editing, I cannot help but hope that CRISPR remains an accessible and valuable tool for citizen scientists who view it with the respect that such a powerful technology deserves. The majority of uses for CRISPR technology are benign and should not be feared. The specific situations in which CRISPR use raises ethical and environmental questions should be conveyed in a transparent manner to the public who will be impacted, and community labs can help ensure that we are all engaged in the shepherding of this new and exciting technology.

REFERENCES

DiCarlo, James E., Julie E. Norville, Prashant Mali, Xavier Rios, John Aach, and George M. Church. 2013. "Genome Engineering in Saccharomyces Cerevisiae using CRISPR-Cas Systems." *Nucleic Acids Research* 41(7): 4336–43.

Global Gene Editing Regulation Tracker. 2023. https://crispr-gene-editing-regs-tracker .geneticliteracyproject.org/united-states-gene-drives/#:~:text=There%20is%20no%20 clear%20gene%20drive%20regulation%20in%20the%20United%20States.

Jinek, Martin, Krzysztof Chylinski, Ines Fonfara, Michael Hauer, Jennifer A. Doudna, and Emmanuelle Charpentier. 2012. "A Programmable Dual-RNA–Guided DNA Endonuclease in Adaptive Bacterial Immunity." *Science* 337 (6096): 816–21.

Jorgensen, Ellen, and Daniel Grushkin. 2011. "Engage With, Don't Fear, Community Labs." *Nature Medicine* 17: 411.

Kahn, Jennifer. 2016. "Gene Editing Can Now Change an Entire Species—Forever." TED talk, February 2016. https://www.ted.com/talks/jennifer_kahn_gene_editing_can _now_change_an_entire_species_forever?language=en.

Lander, Eric. 2015. "The Heroes of CRISPR." *Cell* 164(1–2): 18–28.

Lee, Stephanie M. 2017. "This Guy Says He's the First Person to Attempt Editing His DNA with CRISPR." *BuzzFeed News*, October 14, 2017. https://www.buzzfeednews.com /article/stephaniemlee/this-biohacker-wants-to-edit-his-own-dna#.evELlvD9p.

Matthews, Dylan. 2018. "A Genetically Modified Organism Could End Malaria and Save Millions of Lives—If We Decide to Use It." *Vox*, September 26, 2018. https://www.vox .com/science-and-health/2018/5/31/17344406/crispr-mosquito-malaria-gene-drive -editing-target-africa-regulation-gmo.

Maxmen, Amy. 2015. "Easy DNA Editing Will Remake the World. Buckle Up." *WIRED*, August 2015. https://www.wired.com/2015/07/crispr-dna-editing-2/.

Normile, Dennis. 2018. "CRISPR Bombshell: Chinese Researcher Claims to Have Created Gene-Edited Twins. Reports Trigger Shocked Reactions and Calls for Regulation around the World." *ScienceInsider,* November 26, 2018. https://www.science.org/content/article /crispr-bombshell-chinese-researcher-claims-have-created-gene-edited-twins.

Reardon, Sara. 2015. "US Science Academies Take on Human-Genome Editing." *Nature*, May 18, 2015. https://www.nature.com/articles/nature.2015.17581.

Scheifele, Lisa Z., and Thomas Burkett. 2016. "The First Three Years of a Community Lab: Lessons Learned and Ways Forward." *J Microbiol Biol Educ.* 17(1): 81–85.

Wadhwa, Vivek. 2015. "Why There's an Urgent Need for a Moratorium on Gene Editing." *Washington Post,* September 8, 2015. https://www.washingtonpost.com/news /innovations/wp/2015/09/08/why-theres-an-urgent-need-for-a-moratorium-on-gene -editing/.

Waltz, Emily. 2016. "Gene-Edited CRISPR Mushroom Escapes US Regulation." *Nature* 532:2 93.

Wang, Joy Y., and Jennifer A. Doudna. 2023. "CRISPR Technology: A Decade of Genome Editing is Only the Beginning." *Science* 379 (6629): 251.

Zimmer, Carl. 2012. "Amateurs Are New Fear in Creating Mutant Virus." *New York Times*, March 6, 2012. https://www.nytimes.com/2012/03/06/health/amateur-biologists-are -new-fear-in-making-a-mutant-flu-virus.html?searchResultPosition=5.

5

Welcome to the CRISPR Zoo

Marcus Schultz-Bergin

CRISPR gene editing has been hailed as a revolutionary technology, igniting optimism in our ability to control and influence the genetics of all life on earth (Ledford 2015; Mariscal and Petropanagos 2016; Specter 2015; Zhang and Zhou 2014). CRISPR has made it both possible and practical, now more than ever, to manipulate the genomes of a far greater number and variety of nonhuman animal species (Reardon 2016). While past genetic engineering techniques were often limited to small mammals (especially mice), CRISPR has already shown itself to be much more useful in modifying the genomes of larger animals, including large mammals and nonhuman primates. In view of this increasing use of CRISPR on nonhuman animals, we need to reconsider the ethics of animal genetic engineering. Focusing on the increasing interest in editing agricultural animals, we provide a compelling argument for adopting an animal-focused lens through which to ensure animal welfare and assess the ethics of all CRISPR-based projects.

Gene Editing: Past and Present

CRISPR gene editing is generally considered more precise, versatile, and effective than past genetic engineering techniques. It is also substantially less expensive. We can use the differences in effectiveness and cost to compare CRISPR with past gene-editing techniques to determine whether CRISPR is a more ethical technology for editing nonhuman animal genomes.

One of the most common problems with genetic engineering, and a root cause of many of the major animal welfare issues with past techniques, is the prevalence of off-target edits. Classic genetic engineering techniques have involved the random insertion of genetic material into cells, which resulted in some cells not being altered while others were altered in unintended ways

(Cowan 2015; Ledford 2015). For example, the Beltsville hogs were created in 1985 to speed up growth by introducing human growth hormone (Fox 1990). While some hogs did carry human growth hormone, most also suffered from ulcers and a compromised immune system, likely the result of unintended changes to their gene sequences. Importantly, while not all off-target edits result in deleterious phenotypic traits, many do, and their overall prevalence has limited the application of past genetic engineering techniques.

The greater precision and effectiveness of CRISPR suggests a general solution. Because CRISPR is a targeted gene-editing system, it is more likely to generate the intended genetic changes and avoid off-target results (Carroll and Charo 2015; Zhang and Zhou 2014). CRISPR is, therefore, less likely to produce unintended negative effects that impinge on an animal's physical, psychological, and behavioral well-being or ability to engage in species-typical behaviors (Schultz-Bergin 2017; 2018). CRISPR is still not immune to off-target edits, and there is some concern that the type of off-target edits it may cause, even if less common, may be more deleterious than previous gene-editing technologies (Schaefer et al. 2017). CRISPR's greater efficiency, then, is a double-edged sword. The ability of CRISPR to edit a greater number of cells more efficiently than past techniques applies to both the intended edits and potential off-target mutations (Frock 2015). Thus, while there is a lesser risk of off-target edits in any application of CRISPR, there is potential for greater harm when they do occur.

This potential for greater harm, despite greater precision and efficiency, comes into sharper focus when we consider versatility and cost. With the availability of CRISPR, there has been a significant increase in the types of animals subjected to gene editing as well as the number of experiments being carried out. This expansion of the *CRISPR Zoo* means, first, that even if each individual animal produced by gene editing is less likely to suffer from deleterious welfare effects, gene-edited animals as a population may see a general increase in harm. Individual-level and population-level evaluations may diverge and ethical gains in one area—a reduction in off-target edits—may be offset by ethical losses in another—the increasing number of animal experiments. Even if CRISPR is precise and effective, it is certainly not perfect, and at least some of the resulting animals will suffer.

Greater precision and efficiency are benefits to animal welfare (if they are at all) only when the goal of the edits is, itself, welfare promoting or welfare neutral. But the history of both selective breeding and genetic engineering illustrates that our goals are typically at odds with the animal's welfare

(Rollin 1996; Thompson 1997). The plight of the Beltsville hogs is once again illustrative, for many of them also suffered from weak bones and arthritis precisely because they fattened more quickly, which was the goal of the project. We see something similar with broiler chickens that were selectively bred for quicker growth (Rollin 2003). The CRISPR revolution could simply mean that *more* animals will suffer the ill effects of our march for greater productivity or other human benefit (Schultz-Bergin 2018).

Editing for Animal Welfare

Experiments to improve productivity, even at the expense of animal welfare, will continue, but CRISPR's low cost and versatility has also opened the possibility of projects aimed at improving animal welfare. These existing and proposed projects can be broadly placed into three categories: decreasing suffering, improving health, and reducing cognition.

Our modern agricultural practices cause substantial suffering and death for animals. For instance, the egg and dairy industries require only female animals. What this means, in practice, is that 7 billion one-day-old male chicks are killed every year while male calves are either killed immediately or sold off to the veal industry to live noticeably short lives chained in small crates to avoid building muscle (Krautwald-Junghanns et al. 2018; Levitt 2018). Similar practices exist in the beef industry, where steers are preferred. In the hope of improving this situation, researchers have conducted or proposed experiments that would create intersex animals that are phenotypically the preferred sex, regardless of their chromosomal makeup (Molteni 2020). In the bovine case, this meant adding the *SRY* gene, which instructs mammalian embryos to develop male traits, to the X chromosome of embryos. Although these projects would provide a productivity benefit to the industry, they are also supported as a means of reducing existing suffering and death (Leenstra et al. 2011; Shriver and McConnachie 2018).

In a similar vein, it is currently widespread practice to dehorn cattle, blunt the beaks of chickens, and dock the tails of pigs (American Veterinary Medical Association 2010; Faulkner and Weary 2000; Fulwider et al. 2008; Puppe et al. 2005). Each of these practices is nominally justified by concern for the safety of the animals and humans, but they are also usually done without anesthesia or analgesia and result in acute and chronic suffering. Thus, some CRISPR projects are aimed at editing cows to ensure that they are born hornless and editing chickens to have shorter, blunter beaks (Bhullar et al. 2015;

Carlson et al. 2016). If successful, the idea is that such edits would eliminate the putative need to perform these painful procedures.

Some researchers have also proposed editing animals to improve their health and make them more adaptive to changing environments. Much of this work focuses on disease resistance, including editing pigs to confer tolerance to porcine reproductive and respiratory syndrome (PRRS), a devastating disease for which there is no vaccine (Reiner 2016; Whitworth et al. 2016). Similarly, some have proposed interventions to improve the udder health of dairy cows, because the current practice of overmilking often results in ailments such as mastitis (Eriksson et al. 2018). Finally, due to both changing climate and increasing demand for animal products, some researchers have sought to change animals to better fit current or new environments. For instance, researchers in Israel have produced "naked chickens" that are less sensitive to heat and can withstand hotter climates, such as in the Middle East and Southeast Asia, where demand for chickens is on the rise (Bennet 2002).

A final, more speculative, set of proposals centers on creating *diminished* animals that partially or wholly lack sentience, the conscious capacity to experience pain and pleasure (Schultz-Bergin 2014; Shriver 2009; Shriver and McConnachie 2018; Thompson 2008). In less extreme cases, this deficit may take the form of conscious but numb cows who are more amenable to walking to slaughter (Fox 2020). In more extreme cases, the result may be "headless chickens" or "football birds" that lack consciousness, effectively becoming growing meat sacks (Thompson 2008). In either case, but especially the latter, the basic animal-focused justification is that these animals already experience substantial suffering, we already treat them as nonconscious resources, and we are unlikely to radically change our agricultural practices to reduce their suffering in other ways (Schultz-Bergin 2014; Shriver 2009). Thus, we can eliminate their capacity to suffer and a significant amount of aggregate suffering.

The importance of considering these three proposals is that they potentially counter the general trend in animal genetic engineering of focusing on goals clearly at odds with animal welfare, like increasing meat, milk, or egg production at the expense of safety or suffering. Unlike editing to improve productivity or to enhance medical research, it is possible, at least in principle, that these projects can enhance animal welfare. Given that the possibility of these projects is unique to CRISPR, focusing on them allows us to narrow our ethical assessment to what are arguably new issues, rather than rehashing the classic genetic engineering question: is it ethical to edit animals for our own pleasure or profit?

Editing Animals: An Ethical Assessment

One method for assessing proposals to use CRISPR on animals is to compare them to the status quo: does using CRISPR improve animal welfare compared to our current practices? Proposals to improve animal health with CRISPR do, in fact, compare favorably by making animals less prone to ill health effects that are currently quite common. Proposals aimed at decreasing suffering also fare well, although there are already alternatives to current practice that could similarly do well but may increase cost (Ishii 2017). For instance, it is possible to perform procedures, such as horn removal, under anesthesia and to provide analgesia to manage lingering pain. Alternatively, horn covers can be used to render the removal of horns unnecessary. These changes can be made at the level of the individual animal and do not require the systematic change of animal agriculture.

Proposals to create diminished animals are a bit more difficult to assess in comparison to the status quo, especially the more extreme options that eliminate sentience altogether. Improving animal health and eliminating the need for painful procedures clearly benefit the specific modified animals. But eliminating sentience effectively means creating entities that no longer react to stimuli and become food-production machines. Though these modified animals may not be suffering pain, certainly an ethical good, they could be considered "ethically dead": they would no longer matter from a moral perspective. Although there is philosophical disagreement over what imbues a being with moral status, a family of widely held views consider sentience a necessary (if not sufficient) feature for moral status. If we create beings that lack sentience, then we may be creating beings that lack moral status and, as a result, are "dead to us" from a moral perspective: their interests matter no more than those of a dead animal rotting on the side of the road.

More broadly, we can note the conflict between the individual and population perspectives: to the extent that creating diminished animals *improves* animal welfare at all, it does so by reducing population-level suffering. This effect contrasts with those of other types of proposals that would improve welfare on both levels. The proposals that aim for less extreme diminishment, however, are like those that would ensure the animals are constantly drugged: they would be able to feel but with overall reduced sentience. The result would certainly be less suffering, even at the individual level, but also less capacity for pleasure. If we accept that the status quo provides (almost) no opportunity for animal pleasure, then this trade-off may be morally justified (Shriver and McConnachie 2018).

Further ethical complexity arises when we consider physical health and behavioral well-being. Pain experiences are often valuable signals of physical harm and of avoiding or reducing that harm. Eliminating or numbing pain experiences may result in animals with substantial physical health issues. Similarly, such animals may be incapable or less likely to engage in species-typical behaviors, thus diminishing that aspect of their welfare. Cognitively numbed animals will be less likely to respond to environmental stimuli of any sort, including those that trigger species-typical behaviors. For instance, diminished hens may be less apt to dust bathe or respond to the needs of their young. In some views of animal welfare, the value of these species-typical behaviors is instrumental to improving physical or psychological health. Others consider them to be partly constitutive of an animal's welfare. While the former view may accept a reduction in species-typical behavior if other indicators suggest an improvement in physical and psychological health, the latter view sees the reduction in behaviors as necessarily bad for the animal. While such a view could still justify the trade-off—the animal is made better off physically and psychologically but worse off behaviorally—it holds that there is a loss of a sort.

Nearly all these proposals for using CRISPR on animals will encourage or facilitate the further intensification of animal agriculture, bringing a variety of welfare costs that deserve our ethical consideration (Benz-Schwarzburg and Ferrari 2016; Greenfield 2017). These concerns pertain especially to proposals for improving disease resistance, because disease spread is due to the sheer number of animals kept in cramped conditions. These tight quarters also have other negative welfare effects with ethical implications, such as frustrating the ability to engage in species-typical behaviors like dust-bathing and generating physical ailments like sores and sprained or fractured bones (Rollin 2003). Even as we may make animals more resistant to certain communicable diseases, we may expose them to other harms. Greater use of CRISPR on animals may simply be part of the path of the status quo: we have been selectively breeding animals for millennia. But evidence of pushback against editing agricultural animals for human benefit in certain countries worldwide indicates that using CRISPR may slow, halt, or reverse our progress along that path (Shriver and McConnachie 2018).

Providing a comprehensive ethical judgment about the use of CRISPR on agricultural animals is challenging because of the complex nature of the relationship between human beings and these animals. We can say, however, that if an individual animal were going to be engineered, it is better, in terms of that animal's welfare, that they are engineered with CRISPR than with past techniques. We can also say that CRISPR is likely to provide some general welfare benefit to these animals over our current business as usual. Nonetheless, our history with animal biotechnology should also make us wary that CRISPR, despite its potential for promoting animal welfare, will only further the suffering of agricultural animals, especially where corporate profit is a contributing factor. Were we to make promoting animal welfare our central goal, perhaps CRISPR could be justified. But to do that, we must engage in a variety of other animal welfare reforms that may make using CRISPR largely irrelevant.

REFERENCES

American Veterinary Medical Association. 2010. "Welfare Implications of Beak Trimming." February 7, 2010. https://www.avma.org/resources-tools/literature-reviews/welfare-implications-beak-trimming.

Bennet, James. 2002. "Rehovot Journal; Cluck! Cluck! Chickens in Their Birthday Suits!" *New York Times*, May 24, 2002. https://www.nytimes.com/2002/05/24/world/rehovot-journal-cluck-cluck-chickens-in-their-birthday-suits.html.

Benz-Schwarzburg, Judith, and Arianna Ferrari. 2016. "Super Muscly Pigs: Trading Ethics for Efficiency." *Issues Sci Technol* 32 (3): 29–32.

Bhullar, Bhart-Anjan S., Zachary S. Morris, Elizabeth M. Sefton, Atalay Tok, Masayoshi Tokita, Bumjin Namkoong, Jasmin Camacho, David A. Burnham, and Arhat Abzhanov. 2015. "A Molecular Mechanism for the Origin of a Key Evolutionary Innovation, the Bird Beak and Palate, Revealed by an Integrative Approach to Major Transitions in Vertebrate History." *Evolution* 69 (7): 1665–77. https://doi.org/10.1111/evo.12684.

Carlson, Daniel F., Cheryl A. Lancto, Bin Zang, Eui-Soo Kim, Mark Walton, David Oldeschulte, Christopher Seabury, Tad S. Sonstegard, and Scott C. Fahrenkrug. 2016. "Production of Hornless Dairy Cattle from Genome-Edited Cell Lines." *Nat Biotechnol* 34 (5): 479–81. https://doi.org/10.1038/nbt.3560.

Carroll, Dana, and R. Alta Charo. 2015. "The Societal Opportunities and Challenges of Genome Editing." *Genome Biol* 16 (1). https://doi.org/10.1186/s13059-015-0812-0.

Cowan, Chad. 2015. "Measuring Off-Target Events, Efficiency, and Utility." Presented at the Information-Gathering Meeting for the Planning Committee Organizing the International Summit on Human Gene Editing, Washington, DC.

Eriksson, Susanne, Elisabeth Jonas, Lotta Rydhmer, and Helena Röcklinsberg. 2018. "Invited Review: Breeding and Ethical Perspectives on Genetically Modified and Genome Edited Cattle." *J Dairy Sci* 101 (1): 1–17. https://doi.org/10.3168/jds.2017-12962.

Faulkner, P. M., and D. M. Weary. 2000. "Reducing Pain After Dehorning in Dairy Calves." *J Dairy Sci* 83 (9): 2037–41. https://doi.org/10.3168/jds.S0022-0302(00)75084-3.

Fox, Dov. 2020. "Retracing Liberalism and Remaking Nature: Designer Children, Research Embryos, and Featherless Chickens." *Bioethics* 24 (4): 170–78.

Fox, Michael W. 1990. "Transgenic Animals: Ethical and Animal Welfare Concerns." In *The Bio-Revolution: Cornucopia or Pandora's Box*, edited by Peter Wheale and Ruth McNally. London and Winchester, MA: Pluto Press.

Frock, R. 2015. "Measuring Off-Target Events, Efficiency, and Utility." Presented at the National Academy of Sciences, Washington, DC. https://vimeo.com/142678537.

Fulwider, Wendy K., Temple Grandin, Bernard E. Rollin, Terry E. Engle, Norman L. Dalsted, and W. D. Lamm. 2008. "Survey of Dairy Management Practices on One Hundred Thirteen North Central and Northeastern United States Dairies." *J Dairy Sci* 91 (4): 1686–92. https://doi.org/10.3168/jds.2007-0631.

Greenfield, Andy. 2017. "Editing Mammalian Genomes: Ethical Considerations." *Mamm Genome* 28 (7): 388–93. https://doi.org/10.1007/s00335-017-9702-y.

Ishii, Tetsuya. 2017. "Genome-Edited Livestock: Ethics and Social Acceptance." *Anim Front* 7 (2): 24–32. https://doi.org/10.2527/af.2017.0115.

Krautwald-Junghanns, M.-E., Kerstin Cramer, B. Fischer, Anke Förster, Roberta Galli, Friedrich Kremer, Emmanuel U. Mapesa, et al. 2018. "Current Approaches to Avoid the Culling of Day-Old Male Chicks in the Layer Industry, with Special Reference to Spectroscopic Methods." *Poult Sci* 97 (3): 749–57. https://doi.org/10.3382/ps/pex389.

Ledford, Heidi. 2015. "CRISPR, the Disruptor." *Nature* 522 (7554): 20.

Leenstra, Ferry, G. Munnichs, Volkert Beekman, E. van den Heuvel-Vromans, Lusine H. Aramyan, and Henri Woelders. 2011. "Killing Day-Old Chicks? Public Opinion Regarding Potential Alternatives." *Anim Welf* 20 (1): 37–45.

Levitt, Tom. 2018. "Dairy's 'Dirty Secret': It's Still Cheaper to Kill Male Calves than to Rear Them." *Guardian*, March 26, 2018. http://www.theguardian.com/environment/2018/mar/26/dairy-dirty-secret-its-still-cheaper-to-kill-male-calves-than-to-rear-them.

Mariscal, Carlos, and Angel Petropanagos. 2016. "CRISPR as a Driving Force: The Model T of Biotechnology." *Monash Bioeth Rev* 34 (2): 101–16. https://doi.org/10.1007/s40592-016-0062-2.

Molteni, Megan. 2020. "A Crispr Cow Is Born. It's Definitely a Boy." *Wired*, July 24, 2020. https://www.wired.com/story/a-crispr-calf-is-born-its-definitely-a-boy/.

Puppe, Birger, Peter C. Schön, Armin Tuchscherer, and Gerhard Manteuffel. 2005. "Castration-Induced Vocalisation in Domestic Piglets, Sus Scrofa: Complex and Specific Alterations of the Vocal Quality." *Appl Anim Behav Sci* 95 (1). 67–78. https://doi.org/10.1016/j.applanim.2005.05.001.

Reardon, Sara. 2016. "The CRISPR Zoo." *Nature* 531 (7593): 160–63.

Reiner, Gerald. 2016. "Genetic Resistance - an Alternative for Controlling PRRS?" *Porcine Health Manage* 2 (1): 27. https://doi.org/10.1186/s40813-016-0045-y.

Rollin, Bernard E. 1996. *The Frankenstein Syndrome: Ethical and Social Issues in the Genetic Engineering of Animals*. New York: Cambridge University Press.

Rollin, Bernard E. 2003. *Farm Animal Welfare: Social, Bioethical, and Research Issues*. Malden, MA: Wiley-Blackwell.

Schaefer, Kellie A., Wen-Hsuan Wu, Diana F. Colgan, Stephen H. Tsang, Alexander G. Bassuk, and Vinit B. Mahajan. 2017. "Unexpected Mutations after CRISPR-Cas9 Editing in Vivo." *Nat Methods* 14 (6): 547–48.

Schultz-Bergin, Marcus. 2014. "Making Better Sense of Animal Disenhancement: A Reply to Henschke." *NanoEthics* 8 (1): 101–9.

Schultz-Bergin, Marcus. 2017. "The Dignity of Diminished Animals: Species Norms and Engineering to Improve Welfare." *Ethical Theory Moral Pract* 20 (4): 843–56.

Schultz-Bergin, Marcus. 2018. "Is CRISPR an Ethical Game Changer?" *J Agric Environ Ethics* 31 (2): 219–38. https://doi.org/10.1007/s10806-018-9721-z.

Shriver, Adam. 2009. "Knocking Out Pain in Livestock: Can Technology Succeed Where Morality Has Stalled?" *Neuroethics* 2 (3): 115–24.

Shriver, Adam, and Emilie McConnachie. 2018. "Genetically Modifying Livestock for Improved Welfare: A Path Forward." *J Agric Environ Ethics* 31 (2): 161–80. https://doi.org/10.1007/s10806-018-9719-6.

Specter, Michael. 2015. "The Gene Hackers." *New Yorker*, November 8, 2015. https://www.newyorker.com/magazine/2015/11/16/the-gene-hackers.

Thompson, Paul B. 1997. "Ethics and the Genetic Engineering of Food Animals." *J Agric Environ Ethics* 10 (1): 1–23.

Thompson, Paul B. 2008. "The Opposite of Human Enhancement: Nanotechnology and the Blind Chicken Problem." *NanoEthics* 2 (3): 305–16.

Whitworth, Kristin M., Raymond R. R. Rowland, Catherine L. Ewen, Benjamin R. Trible, Maureen A. Kerrigan, Ada G. Cino-Ozuna, Melissa S. Samuel, et al. 2016. "Gene-Edited Pigs Are Protected from Porcine Reproductive and Respiratory Syndrome Virus." *Nat Biotechnol* 34 (1): 20–22. https://doi.org/10.1038/nbt.3434.

Zhang, LinLin, and Qi Zhou. 2014. "CRISPR/Cas Technology: A Revolutionary Approach for Genome Engineering." *Sci China Life Sci* 57 (6): 639–40. https://doi.org/10.1007/s11427-014-4670-x.

III PERSONAL PERSPECTIVES

6

Who Goes First?

Carol Padden and
Jacqueline Humphries

Six years ago, the US National Academies of Sciences, Engineering, and Medicine (NASEM 2017) released a report drafted by an international committee regarding the use of gene editing in humans. Once a tedious and expensive process, gene editing has now become more feasible using rapidly maturing CRISPR technology, making the issue of its use more urgent. The committee cites general support for somatic nonheritable gene editing to treat existing serious diseases, but more controversial is heritable genome editing (HGE), which affects descendants. HGE is tightly controlled in the United States, but around the world, scientists are pushing against the regulatory barrier. The committee includes a recommendation that the pursuit of HGE should be limited to "serious diseases and conditions" they define elsewhere in the report as fatal or debilitating—though what is considered debilitating is open to debate. Their recommendations were echoed in a statement signed by the Organizing Committee of the Second International Summit on Human Genome Editing that "we continue to believe that proceeding with any clinical use of germline editing remains irresponsible at this time" (NASEM 2019, 132). The statement confirms what most biologists believe—that there are significant technical challenges remaining before HGE can be used to treat individuals, aside from the fact that ethical and clinical standards for implementation have yet to be developed.

Deaf People and CRISPR Germline Editing

For the authors of this essay, the topic of gene editing is close to home professionally and personally. We live with genetic deafness, and the languages of our home are American Sign Language (ASL) and English. Carol Padden is deaf, as are her parents and grandparents. ASL has been used in her home for at least three generations. Carol's daughter, Jacqueline Humphries, is hearing with two deaf parents (and grandparents) and is bilingual in ASL and English. In our family, we often discuss the scientific and ethical case for gene editing, as well as the cultural and social case for diversity.

Fundamental to our discussions about gene editing is the view that genetic diversity and cultural diversity are constitutive of each other. Practicing good science and medicine means understanding that humans are constituted by their cultural and social practices. Culture is not a byproduct of human evolution, but phylogenetically, culture coemerged with human cognition and language environments. Coller (2019) describes mutations as not "errors . . . but rather an intrinsic and important aspect of our species' evolutionary success" (293).

Genetic deafness includes various genetic conditions involving up to 200 to 250 genes. The most frequent, accounting for over 80% of inherited deafness, are autosomal recessive and nonsyndromic, meaning that there are no obvious conditions other than deafness. These conditions are not fatal, but some may argue that the inability to hear is debilitating. Because CRISPR could eliminate at least some of the alleles that confer deafness, it is crucial to debate openly and seriously about which types of conditions might qualify for HGE, if any. In a review of the debate around gene editing, Coller (2019) cites members of the deaf community as "contending . . . that deafness is not a disability, but rather a manifestation of human diversity and that protecting and preserving the deaf community's culture is important" (295). Shearer, Hildebrand, and Smith (2017), in an overview of hereditary hearing loss, include the term *Deaf culture* in their glossary, explaining that there are "members of the Deaf community in the US" who do not regard deafness as "a pathology or disease to be treated or cured" (2). Their mention of cultural views of deaf people in medical literature is novel, considering that current medical texts describe deafness, and broadly hearing loss, as a pathology. The widespread acceptance and use of cochlear implants in young deaf children show that the prevailing medical and public view of deafness is that it should be repaired as early as possible and eventually eliminated.

Hearing loss is described as the most frequent sensory impairment appearing in newborns (Atik et al. 2015). In Europe and North America, most cases of deafness result from genetic inheritance. In other parts of the world, hearing loss more often results from acquired illnesses, but genetic causes account for many cases of hearing loss across all societies (Shearer, Hildebrand, and Smith 2017). One common genetic cause involves mutations of the gene *GJB2*, which codes for the gap junction protein connexin 26. These mutations affect passage of ions through gap junctions and interfere with the transmission of auditory signals within the cochlea, resulting in hearing loss (Kemperman, Hoefsloot, and Cremers 2002). While multiple mutations in connexin 26 can result in deafness, one of the most common variants is known as 35delG. This variant is most prevalent among individuals of European descent or from the Middle East and Turkey, and it is also found to a lesser extent in those of Asian descent (Gasparini et al. 2000; Mahdieh and Rabbani 2009; Tekin et al. 2003). A substantial proportion of individuals with autosomal recessive nonsyndromic hearing loss have mutations in connexin 26. Because it is recessive, an individual must inherit two copies of the mutation, one from each parent. Their parents may be deaf themselves, or each hearing parent may be a carrier of just one copy of the mutated variant. Researchers have observed high carrier frequencies of up to 1 in 35 for the 35delG variant of connexin 26 in certain populations (Gasparini et al. 2000). The question, then, is why carriers of mutations in connexin 26 are so common and widespread in various parts of the world.

The answer may lie in a phenomenon called *heterozygote advantage*, where having one wild type—unmutated—copy and one mutated copy of a gene can present an advantage in certain conditions. A widely discussed example is resistance to malaria among carriers of sickle cell anemia (Luzzatto 2012). In the case of genetic deafness, Meyer and colleagues (2002) observed that carriers or homozygotes of a certain mutation in *GJB2* had thicker epidermis than family members without a mutant allele. Furthermore, individuals who were homozygous for the mutant allele had increased ion concentrations in their sweat. These changes might influence susceptibility to pathogens, providing a selective advantage to this mutation. Variation in the human gene pool is valuable, and the choice to remove such variation must be balanced against the severity of the condition it is intended to prevent. At present, there is not enough research on connexin-related conditions and heterozygote advantage—ironically, because deafness is not considered to be so debilitating that it warrants sizable investment in this kind of research. In general, we think any potential of a heterozygote advantage in a particular human population should give scientists serious pause before they explore germline editing to remove or reduce that advantage.

As stated, the immediate challenge in germline editing of deafness is the considerable number of genes involved in hearing. Further, individuals can carry different mutations in the same gene. As Atik and colleagues (2015) note, "this overwhelming genetic and clinical heterogeneity makes the identification of genetic etiology challenging, time consuming, and expensive" (2). Moving beyond identification to targeted editing of mutations is a challenge, because any gene-editing procedure must protect against the chance of off-target edits, where edits are introduced in unintended places in the genome. Currently, it is difficult to develop a process that can reliably edit a single allele with no off-target effects, much less the more than 1,000 mutations associated with hearing loss.

But the enormity of the challenge is likely to do little to dissuade medicine from its goal of eradicating deafness. An abundance of nonheritable gene therapy studies is directed at the cochlea of mice carrying mutant alleles that cause hearing loss in humans, including connexin 26 (Zhang et al. 2018). We believe that clinical attempts to carry out germline editing of certain forms of genetic deafness will take place in the future. The question is—how soon?

The *New Scientist* reported that a Russian biologist had announced plans to proceed with germline editing on behalf of five deaf couples, all of whom have recessive deafness caused by mutations in the *GJB2* gene (Le Page 2017). The article quoted a bioethicist, Julian Savulescu, as warning against such an attempt: "The first human trials should start with embryos or infants with nothing to lose, with fatal conditions . . . you should not be starting with an embryo which stands to lead a pretty normal life." While we cringe at any reference to a human infant "with nothing to lose," as the right justification for experimentation, more to the point is the question: why is this Russian biologist choosing to start human genome–editing trials with deaf couples?

If connexin-26 recessive deafness is not fatal or debilitating, why might a scientist defy the recommendations of learned professional societies and committees and proceed with clinical studies involving deaf individuals? We think it is because deaf people, particularly those who enjoy the benefits of multimodal language and are active participants in self-defining communities and cultures, are ideal clinical subjects. First, those who have autosomal nonsyndromic recessive deafness are likely to be healthy and free of other inheritable diseases, unlike individuals with sickle cell anemia or cystic fibrosis (CF). Second, they are also more likely to marry other deaf people who share the same facility for communication, and thus they may end up with spouses who have the same recessive mutations. In their offspring, only one copy of the gene would have to be edited to prevent deafness. Third, deaf individuals who reach adulthood are often active members of society, and in developed parts of the world, they are generally educated and can give informed consent regarding clinical trials. Finally, adult couples with recessive deafness of the *GBJ2* gene are not at all rare—indeed, in many parts of the world, there are good numbers of them for a substantial clinical trial.

As a comparison scenario, consider CF, which is a serious condition. If germline editing were to become feasible, CF could be among those diseases to attract support for such an intervention, because many individuals with CF die before age 30, though non-gene-editing breakthroughs have recently extended life expectancy. Those with CF who reach adulthood are still actively discouraged from marrying each other and having families because of the substantial risk of exposure to bacteria that might be resistant to the armamentarium of antibiotics they must take to treat pulmonary infections. Further, men with CF are commonly infertile because CF affects the vas deferens. Compared to deafness, CF would not provide a readily available population of homozygous couples for clinical studies, so editing for CF is logistically far more difficult.

Recommendations from learned societies to wait for further research on germline editing may not be enough to dissuade scientists who believe it is a valuable tool for human medicine. Generations of medical professionals throughout history have decided that deaf children and adults need intervention. Tragically, they have been subject to sterilization and experimental surgery with little or no benefit. Now, potentially, they will be subjects for a new generation of CRISPR studies.

"CRISPRing" Disabilities versus Embracing Human Diversity

Some humanities scholars describe the rush to repair and eradicate disability, including deafness, as resulting from an "ableist logic" which affirms "a network of beliefs, processes and practices that produces a particular kind of self and body . . . that is projected as perfect, species-typical and therefore essential and fully human." In this view, disability "is cast as a diminished state of being human" (Campbell 2012, 213). It is hard to counter these beliefs, not just in medicine but in nearly every institution of modern life: education, the labor force, and public policy. Indeed, as disability scholars argue, the very fact of modernity necessitates a belief in repair and augmentation of the human body toward an ideal of perfection, even enhanced ability. The ubiquity of these ideas has been the focus of the emerging field of critical disability studies, whose project has been to counter the prejudice that has "negative social consequences for disabled people," as well as to "dis-figure the non-disabled" in their impossible desire to attain what cannot be achieved (Hughes, Goodley, and Davis 2012, 313).

The breadth of the ways that prejudice enters into the lives of deaf people, particularly about their sign language, is difficult to contain and impossible to review here (Humphries et al. 2017). Instead, we offer two compelling ex-amples of deafness and the history of deaf people as expressions of cultural diversity, not contracting but expanding the abilities and capacities of human beings. Our examples come from those practices that are familiar to us in our lives—that is, how we live among deaf people and work with hearing people.

Sign language and signers have likely always been a part of human so-ciety, or at the very least when language became possible in the species. Modern-day sign language research is traced to the middle of the last century, when a small group of linguists became interested in whether human language could be found in a modality other than speech (Klima and Bellugi 1979; Stokoe, Casterline, and Croneberg 1965). Most of this work began with the easily accessible sign languages in the United States and Europe (at least those sign languages near colleges and universities where this research was undertaken). There are large communities of signers in both regions of the world, and they descend from durable institutions tied to educational sys-tems for deaf children. A recent estimate is that there are about 1,000,000 primary users of ASL in North America, which would rank it as among the larger non-English languages in the United States (Mitchell and Young 2023). Beyond this number, there are many more who have learned it as a second language and are fluent. National sign languages, or sign languages broadly used within countries, exist worldwide. (An inventory of these sign lan-guages can be found in glottolog.org.)

Recently, sign language linguists have turned their attention to smaller and exceedingly small communities of signers where new sign languages have emerged in recent time, within the current or last two or three generations (Zeshan and de Vos 2012). Many of these communities involve the recent ap-pearance of deafness resulting from intermarriage between individuals who are carriers of autosomal-dominant or autosomal-recessive deafness. The sci-entific value of such communities is that they offer the opportunity to ob-serve how human language emerges in as few as one to two generations of signers. These emergent forms allow scientists to explore how the founda-tional elements of language—words—are combined to form propositions and sentences, allowing complex discourse (Kastner et al. 2014; Lanesman and Meir 2012; Sandler et al. 2005). The new interest in emerging sign language communities is part of a broader trend in science to study the dimension of time or emergent phenomena to understand underlying principles and properties.

Several studies on small-village sign languages have been carried out in communities around the world where there is genetic deafness, including the Middle East, Turkey, Bali, Ghana, Algeria, Mali, Mexico, and Thailand (Aronoff et al. 2008; de Vos 2015; Ergin et al. 2018; Hou 2016; Lanesman and Meir 2012; Nonaka 2009; Nyst 2000, 2015). In one of the more well-known cases in southern Israel in a Bedouin community, the sign language used in the community is traced to a family with deaf siblings in the 1930s who shared a mutation in the connexin 26 gene (Meir et al. 2010; Scott et al. 1998). Today, there are at least 150 deaf adults and children in a community of 6,000; in addition to deaf signers, there are at least three times as many hearing family members who also use that particular sign language with their parents, children, siblings, and other relatives throughout the community. Other communities that have been studied are smaller, and some consist only of immediate family members, but in all these cases, the common observation is how readily humans living together in a community will adapt to using language in a modality other than speech.

This fact is not surprising in the view of psychologists and linguists who study the human use of gesture in everyday life (Goldin-Meadow 2005; Kendon 2000; Padden et al. 2015). Gesture is often thought of as an accompaniment to speech, but it is now understood as integral to human language and a way to extend audible limitations of speech. To say that something is "over there" requires pointing. To describe detailed shapes and physical properties, gesture is indispensable. Goldin-Meadow (2006) has argued that the gesture not only communicates, but also organizes and directs thought during speech. The deep relationship between gesture and thought has led some to argue that the emergence of the symbolic human being was not by speech alone but by gestural abilities as well (Kendon 2000).

We argue that the many cases of spontaneous emergence of sign language in modern-day communities of hearing and deaf people around the world demonstrate that humans are naturally multimodal. Speech appearing in every human community does not nullify the potential for multimodality, nor does it confirm that humans are exclusively unimodal for speech. Humans around the world speak, gesture, and sign. They will participate in the creation of complex gestures, leading to a new sign language over time if the conditions for sustained use of a sign language persist long enough for it to flourish. In this sense, sign language is not a marker of failure, used only by those who cannot use speech. It is an indicator of an adaptable ability in humans to create and use languages across multiple modalities for different purposes.

The argument for multimodality as the prevailing characteristic of human beings need not be limited to examples of sign language communities. It extends to all of us in our technologically infused lives. It is startling to deaf people who have grown up using subtitles and captions to watch hearing people also using captions in public places and at home. Captions are now easily available, generated either by human transcribers or by automation, for videos across all media platforms. At first, closed captions were required by law in the United States to enable accessibility to the public airwaves by the disabled. Now, nearly 40 years after they first appeared on television for spoken English, they are used by hearing people as well. Americans turn on captions while watching British television, or while jogging on the treadmill. Giant screens in Times Square feature captions for televised entertainment. Video clips on a website are often captioned so that viewers do not need to listen to them while scrolling down a page; they can be quickly read in the short length of time it takes to scan the page. Texting and email have taken over a large part of the interaction that once took place on a telephone. To be sure, the expansion of audible communication, from podcasts to readily available music in all media has been remarkable, but the rise of visual and textual communication has been just as explosive. It can be said that building technological systems for the disabled has resulted in expanded abilities for the abled.

As we once noted in a book about the modern deaf community in the United States, we live in times of enormous possibilities but also enduring prejudices (Padden and Humphries 2009). The rapid rise of ASL classes in the United States, not only in universities, but in high schools and even middle and elementary schools, is nothing short of astonishing. Young parents today use sign language vocabulary with their prelingual hearing infants as a route for early language before their speech becomes intelligible.

Setting Goals—and Boundaries—for CRISPR

If the goal of medicine is to heal and improve lives, we harbor no illusion that it will ever view deafness as compatible with these goals, at least not in the present. Though we are "the public" who should be heard in matters such as gene editing, our views—that deafness is one of many variations of being human and should be valued, not pathologized—clash with those of many medical professionals and probably many geneticists. Too many parents, doctors, teachers, and other professionals do not view deafness the way we do; the public is more likely to include them than us. Coller (2019) acknowledges as much: "healthcare professionals tend to overestimate the impact of some disabilities on life satisfaction of children and their families" (295).

We stand in alliance with various organizations and scientists who work at the front lines of ethical use of scientific knowledge about gene editing. They include orthodox Jewish communities that have instituted selected genetic testing they believe is compatible with their cultural practices, though there is disagreement among them about which genetic conditions should be tested and how to avoid stigmatization of carriers (Ekstein and Katzenstein 2001; Raz and Vizner 2008). Another debate revolves around consanguinity in communities around the world where marriage between close relatives is commonly practiced despite the risk in increasing the incidence of certain genetic conditions (Bittles 2002). As members of such communities have explained, the overriding cultural consideration of consanguinity is the maintenance of close ties between families and within related families. The counterbalancing force in the gene-editing debate will have to be "the public," which must include those of us who live the condition, in cultures around the world.

Complicating matters is that from communities of deaf people to those of orthodox Jews and couples who are closely related, there are individuals who will support gene editing. There are deaf couples who want to choose to have hearing children. Some orthodox Jews want more development of genetic testing for rare conditions known to affect members of their community. A suitable alternative to germline editing for parents might be to adopt from a fertility clinic a leftover embryo that does not have the mutation (NASEM 2017, 113). But because many parents desire to conceive genetically related children, this readily available alternative is not given the consideration it deserves.

Viotti and coauthors (2019) sidestep the culture debate and focus instead on estimating the clinical demand for HGE. They argue that because there is no safe, risk-free way to carry out germline editing, there is no clinical demand at present. Furthermore, even if there were soon a viable way to do the procedure, the authors find that there are very few cases that could qualify for it. Their estimate is "fewer than a dozen clinical cases . . . born per year in the United States that might hypothetically benefit from germline gene editing" (308). In the case of genetic deafness, which the authors describe as a "less severe hereditary condition," they say the availability of cochlear-implant technology currently "might obviate the need for germline genetic editing and its associated risks" (311). They remain careful, however, not to rule out editing as a procedure in the future.

As we conclude, we return to our abiding argument: there is an overriding concern that the debate around who should first benefit from new medical breakthroughs favors expediency and urgency of scientific discovery. Deaf individuals may find themselves first in line for germline editing because their autosomal nonsyndromic recessive condition offers ideal clinical opportunities compared to those with more serious or debilitating diseases. To practice good science and good public policy, however, we urge caution and care to understand how to promote and not diminish genetic and cultural diversity, both of which offer our best possibilities for living well into the future.

REFERENCES

Aronoff, Mark, Irit Meir, Carol Padden, and Wendy Sandler. 2008. "The Roots of Linguistic Organization in a New Language." *Interaction Stud* 9 (1): 133–53. https://doi.org/10.1075/is.9.1.10aro.

Atik, Tahir, Huseyin Onay, Ayca Aykut, Guney Bademci, Tayfun Kirazli, Mustafa Tekin, and Ferda Ozkinay. 2015. "Comprehensive Analysis of Deafness Genes in Families with Autosomal Recessive Nonsyndromic Hearing Loss." *PLoS One* 10 (11): e0142154. https://doi.org/10.1371/journal.pone.0142154.

Bittles, Alan Holland. 2002. "The Impact of Consanguinity on the Indian Population." *Indian J Hum Genet* 8 (2): 45–51.

Campbell, Fiona Kumari. 2012. "Stalking Ableism: Using Disability to Expose 'Abled' Narcissism." In *Disability and Social Theory: New Developments and Directions*, edited by Dan Goodley, Bill Hughes, and Lennard Davis, 212–30. New York: Palgrave Macmillan.

Coller, Barry S. 2019. "Ethics of Human Genome Editing." *Annu Rev Med* 70: 289–305. https://doi.org/10.1146/annurev-med-112717-094629.

de Vos, Connie. 2015. "The Kata Kolok Pointing System: Morphemization and Syntactic Integration." *Top Cogn Sci* 7 (1): 150–68. https://doi.org/10.1111/tops.12124.

Ekstein, Josef, and Howard Katzenstein. 2001. "The Dor Yeshorim Story: Community-Based Carrier Screening for Tay-Sachs Disease." *Tay-Sachs Dis* 44: 297–310.

Ergin, Rabia, Irit Meir, Deniz Ilkbaşaran, Carol Padden, and Ray Jackendoff. 2018. "The Development of Argument Structure in Central Taurus Sign Language." *Sign Lang Stud* 18 (4): 612–39. https://doi.org/10.1353/sls.2018.0018.

Gasparini, Paolo, Raquel Rabionet, Guido Barbujani, Salvatore Melchionda, Michael Petersen, Karen Brønum-Nielsen, Andres Metspalu, et al. 2000. "High Carrier Frequency of the 35delG Deafness Mutation in European Populations." *Eur J Hum Genet* 8 (1): 19–23. https://doi.org/10.1038/sj.ejhg.5200406.

Goldin-Meadow, Susan. 2005. *Hearing Gesture: How Our Hands Help Us Think*. Cambridge: Harvard University Press.

Goldin-Meadow, Susan. 2006. "Talking and Thinking with Our Hands." *Curr Dir Psychol Sci* 15 (1): 34–39.

Hou, Lynn Yong-Shi. 2016. "'Making hands': Family Sign Languages in the San Juan Quiahije Community." PhD diss. University of Texas.

Hughes, Bill, Dan Goodley, and Lennard Davis, eds. 2012. "Conclusion: Disability and Social Theory." In *Disability and Social Theory: New Developments and Directions*, 308–17. New York: Palgrave Macmillan.

Humphries, Tom, Poorna Kushalnagar, Gaurav Mathur, Donna Jo Napoli, Carol Padden, Christian Rathmann, and Scott Smith. 2017. "Discourses of Prejudice in the Professions: The Case of Sign Languages." *J Med Ethics* 43 (9): 648–52.

Kastner, Itamar, Irit Meir, Wendy Sandler, and Svetlana Dachkovsky. 2014. "The Emergence of Embedded Structure: Insights from Kafr Qasem Sign Language." *Front Psychol* 5: 1–15. https://doi.org/10.3389/fpsyg.2014.00525.

Kemperman, Martijn H., Lies H. Hoefsloot, and Cor W. R. J. Cremers. 2002. "Hearing Loss and Connexin 26." *J R Soc Med* 95 (4): 171–77. https://doi.org/10.1177/014107680209500403.

Kendon, Adam. 2000. "Language and Gesture: Unity or Duality." In *Language and Gesture*, edited by David McNeill, 47–63. New York: Cambridge University Press.

Klima, Edward S., and Uursula Bellugi. 1979. *The Signs of Language*. Cambridge: Harvard University Press.

Lanesman, Sara, and Irit Meir. 2012. "The Survival of Algerian Jewish Sign Language alongside Israeli Sign Language in Israel." In *Sign Languages in Village Communities: Anthropological and Linguistic Insights*, edited by Ulrike Zeshan and Connie de Vos, 153–80. Boston: De Gruyter.

Le Page, Michael. 2017. "Exclusive: Five Couples Lined up for CRISPR Babies to Avoid Deafness." *New Scientist*, July 13, 2017. https://www.newscientist.com/article/2208777-exclusive-five-couples-lined- up-for-crispr-babies-to-avoid-deafness/.

Luzzatto, Lucio. 2012. "Sickle Cell Anaemia and Malaria." *Mediterr J Hematol Infect Dis* 4 (1). https://pubmed.ncbi.nlm.nih.gov/23170194/.

Mahdieh, Nejat, and Bahareh Rabbani. 2009. "Statistical Study of 35delG Mutation of GJB2 Gene: A Meta- Analysis of Carrier Frequency." *Int J Audiology* 48 (6): 363–70. https://doi.org/10.1080/14992020802607449.

Meir, Irit, Wendy Sandler, Carol Padden, and Mark Aronoff. 2010. "Emerging Sign Languages." In *Oxford Handbook of Deaf Studies, Language, and Education, Volume 2*, edited by Marc Marschark and Patricia Elizabeth Spencer, 267–80. Oxford: Oxford University Press.

Meyer, Christian G., Geoffrey K. Amedofu, Johanna M. Brandner, Dieter Pohland, Christian Timmann, and Rolf D. Horstmann. 2002. "Selection for Deafness?" *Nat Med* 8 (12): 1332–33. https://doi.org/10.1038/nm1202-1332.

Mitchell, Ross E., and Travas A. Young. 2023. "How Many People Use Sign Language? A National Health Survey-Based Estimate." *Journal of Deaf Studies and Deaf Education* 28 (1): 1–6. https://doi.org/10.1093/deafed/enac031.

National Academies of Sciences, Engineering, and Medicine (NASEM). 2017. *Human Genome Editing: Science, Ethics, and Governance*. Washington, DC: National Academies Press. https://doi.org/10.17226/24623.

National Academies of Sciences, Engineering and Medicine (NASEM). 2019. *Second International Summit on Human Genome Editing: Continuing the Global Discussion: Proceedings of a Workshop—in Brief*. Washington, DC: National Academies Press. https://doi.org/10.17226/25343.

Nonaka, Angela M. 2009. "Estimating Size, Scope, and Membership of the Speech/Sign Communities of Undocumented Indigenous/Village Sign Languages: The Ban Khor Case Study." *Lang Commun* 29 (3): 210–29.

Nyst, Victoria Anna Sophie. 2000. *A Descriptive Analysis of Adamorobe Sign Language (Ghana)*. Utrecht: Netherlands Graduate School of Linguistics.

Nyst, Victoria Anna Sophie. 2015. "The Sign Language Situation in Mali." *Sign Lang Stud* 15 (2): 126–50.

Padden, Carol, and Tom Humphries. 2009. *Inside Deaf Culture*. Cambridge: Harvard University Press.

Padden, Carol, So-One Hwang, Ryan Lepic, and Sharon Seegers. 2015. "Tools for Language: Patterned Iconicity in Sign Language Nouns and Verbs." *Top Cogn Sci* 7 (1): 81–94. https://doi.org/10.1111/tops.12121.

Raz, Aviad, and Yafa Vizner. 2008. "Carrier Matching and Collective Socialization in Community Genetics: Dor Yeshorim and the Reinforcement of Stigma." *Soc Sci Med* 67 (9): 1361–69. https://doi.org/10.1016/j.socscimed.2008.07.011.

Sandler, Wendy, Irit Meir, Carol Padden, and Mark Aronoff. 2005. "The Emergence of Grammar: Systematic Structure in a New Language." *Proc Natl Acad Sci* 102 (7): 2661–65. https://doi.org/10.1073/pnas.0405448102.

Scott, D. A., M. L. Kraft, R. Carmi, A. Ramesh, K. Elbedour, Y. Yairi, C. R. Srikumari Srisailapathy, et al. 1998. "Identification of Mutations in the Connexin 26 Gene that Cause Autosomal Recessive Nonsyndromic Hearing Loss." *Hum Mutat* 11 (5): 387–94. https://doi.org/10.1002/(SICI)1098-1004(1998)11:5<387::AID-HUMU6>3.0.CO;2-8.

Shearer, A. Eliot, Michael S. Hildebrand, and Richard. J. Smith. 2017. "Hereditary Hearing Loss and Deafness Overview." In *Gene Reviews*, edited by Margaret Adam, Holly H. Ardinger, and Roberta A. Pagon, 1993–2019. Seattle: NCBI Bookshelf.

Stokoe, William, Dorothy C. Casterline, and Carl Croneberg. 1965. *A Dictionary of American Sign Language on Linguistic Principles*. Washington, DC: Gallaudet College Press.

Tekin, Mustafa, Türker Duman, Gönül Boğoçlu, Armağan Incesulu, Elif Comak, Inci Ilhan, and Nejat Akar. 2003. "Spectrum of GJB2 Mutations in Turkey Comprises both Caucasian and Oriental Variants: Roles of Parental Consanguinity and Assortative Mating." *Hum Mutat* 21 (5): 552–53. https://doi.org/10.1002/humu.9137.

Viotti, Manuel, Andrea R. Victor, Darren K. Griffin, Jason S. Groob, Alan J. Blake, Christo G. Zouves, and Frank L. Barnes. 2019. "Estimating Demand for Germline Genome Editing: An *In Vitro* Fertilization Clinic Perspective." *CRISPR J* 2 (5): 304–15. https://doi.org/10.1089/crispr.2019.0044.

Zeshan, Ulrike, and Connie de Vos. 2012. *Sign Languages in Village Communities: Anthropological and Linguistic Insights*. Boston: De Gruyter.

Zhang, Wenjuan, Sun Myoung Kim, Wenwen Wang, Cuiyuan Cai, Yong Feng, Weijia Kong, and Xi Lin. 2018. "Cochlear Gene Therapy for Sensorineural Hearing Loss: Current Status and Major Remaining Hurdles for Translational Success." *Front Molec Neurosci* 11: 221. https://doi.org/10.3389/fnmol.2018.00221.

7

Billie Idol

Ethan Weiss

Billie Idol was the name we gave to Ruthie in the hospital in the days immediately after she was born. She had fluorescent white hair, and had she been born to different parents, they might have thought more of it. But both of Ruthie's parents had white blond hair as young children. So, in the late summer and into the fall of 2006, we happily celebrated the arrival of our second child, little blond baby Ruthie "Billie Idol" Weiss.

Like many second-time parents, we doted less. We did not love less, but we were clearly less excited. We were much more exhausted. We also paid a little less attention. Around the time Ruthie turned one month old, my wife, Palmer, started to tell me about things she had noticed that seemed off. By that time, we had been married for four years and had been together for six. But we also had roles. I was the physician and the scientist, so I would be the one who managed the business of our family's health. Having grown up with a neurotic physician father, I reacted by learning to minimize and dismiss things with ease. I was a minimalist, at least when it came to my health or the health of my family.

I will never forget where I sat in our kitchen that fall when Palmer came back from a playdate she had attended with a group of her friends and their new babies. She was not overly concerned, and sometimes I expected that she just wanted me to reassure her. She said she really did think there might be something wrong with Ruthie. In retrospect, I now know that she was not looking for reassurance. She said that Ruthie's eyes did not track, and she noticed it when she picked up her friend Katherine's baby, who locked eyes with her as they smiled together.

I flat out dismissed it. I said something about how normal development is on a spectrum and that Ruthie was just going to take her time doing things her way. Young parents all too often fall into the trap of comparing their children to other children. It almost never ends well.

A few weeks later, I got sick and decided to stay home from work. I distinctly remember that it was a Friday and that I looked forward to hanging out with Ruthie—alone. We did what sick adults and young babies do for most of the day: we slept. Sometime that afternoon, I mustered the energy to change her diaper. I put Ruthie on her back, and as I got everything together, I noticed the slow rhythmic beating of her eyes back and forth like a metronome. It was mesmerizing. But then my mind raced. As I hovered over my daughter, it took but a few seconds to come to grips with the realization that something was, indeed, abnormal.

In my stupor, I left Ruthie on her back on our bed and went to the room that then served as our study. I sat at the computer, opened the Internet browser, navigated to UptoDate.com, and typed in "infant nystagmus." I clicked on the article, and what came back was a list of mostly horrible neurological conditions. Nowhere in the article did I see the word *normal*. But then it hit. My eyes fixated on it: "oculocutaneous albinism." Right then, I knew. I knew that the funny jokes about Billie Idol were not so funny.

I spent a few minutes convincing myself that I was wrong, but that was impossible. I went back to Ruthie, who incidentally smiled the entire time I was gone and has not stopped smiling since. I finally did change her diaper, and then I held her as I thought about how I would tell Palmer when she got home from work later that day. Of course, knowing that it was a Friday afternoon, I also knew that there was little chance we would get confirmation of the diagnosis until Monday at the earliest. I knew the weekend was going to be long. When Palmer finally came home, I asked her to sit down. That is a terrible thing to do, and I have never done it since: there is no more panic-inducing request in the English language. Palmer did not sit down, but instead demanded immediately to know, "what?" Mothers really do always know.

My mind was hazy, and I would have struggled even under the best circumstances. Maybe it was easier this way. Maybe being sick disinhibited me? I told her, "I think I know what is wrong with Ruthie."

"What?!"

I said, "I think she has albinism."

Palmer immediately burst into tears. So did I. Her mind raced with images of evil characters from movies. Albino. Albino. That word. Our housekeeper came into our bedroom to check on us. She spoke almost no English, but she did not need to.

We spent most of the next 48 hours crying. We were terrified. We were sad. We were angry. We managed to schedule an emergency visit to a great pediatric ophthalmologist for Monday afternoon, and we waited. But we knew.

It took Dr. Day about one minute to look at Ruthie and confirm that she had albinism. She did not even have to look in her eyes, but she did. She looked at us and said directly, "Ruthie has albinism."

Our questions were mostly directed at how bad Dr. Day expected her vision to be, and with great prescience, she said she thought Ruthie had some pigment and predicted her visual acuity would be about 20/200: what we could see from 200 feet, Ruthie would need to be 20 feet away from. Would she drive a car? Would she date? Could she go to a normal school? Over the next few weeks, we experienced every emotion there was. Dr. Day arranged for us to talk to one of her former patients with albinism and her mother. The former patient's father, also an ophthalmologist, had diagnosed her with albinism at birth. We went to see a dermatologist. We went to the geneticist where we agreed we would do genetic testing on Ruthie. One of the darkest moments was the day we took her to have her blood drawn for genetic testing. I am quite sure it was the first time she had really cried.

The decision to do genetic testing was reflexive. We did not think hard about why we should or should not do it. We did not think about how the information might change our thinking or Ruthie's management. At that time, we did not consider that we very well might have learned the genetic basis for her albinism long before Ruthie was born or even while Palmer was pregnant. We did not think of the implications such knowledge might have had, especially at a time when we were unprepared to even know what it all meant. Of course, Ruthie's albinism has had a profound impact on me as a person and as a doctor. I do not now encourage my patients to do genetic testing reflexively or just because they can. I encourage them to think hard about what the outcomes might be, what they might mean, and how they might change things in a good or bad way. I encourage them to imagine how the results might affect them. How might it change their lives or their children's lives? What are the potential positive outcomes? What are the risks? I also remind them that we often cannot know the answers to these questions in advance. Overall, I am a different person now. I am a different doctor now.

We also saw a therapist to discuss how we would communicate Ruthie's condition to her older sister, our family, our friends, and the world. We even talked about whether we needed to tell anyone. The discussion with the therapist was important and empowering. We knew how we would handle these questions, but it was important to say them aloud and have the process validated. In retrospect, it was a fast and easy conversation, and we ended up spending the bulk of the hour talking about the challenges of parenting our genetically normal three-year-old.

Later that fall, we received her genetic testing results. Ruthie had inherited one mutated copy of the *OCA2* gene from me and one from Palmer. She was what is called a *compound heterozygote* and was left with two partially functioning copies of the gene, hence albinism.

We eventually decided to tell everyone about Ruthie's albinism. Palmer and I agreed that we would not run from reality. We would embrace it and share it. We wanted her difference to be as normal as it could be, and we knew we had to lead by example. We had to help the world be comfortable with Ruthie. Sharing genetic responsibility may have made that easier. Since we knew that she had inherited a mutant copy of the gene from both of us, we also knew that we each bore 50% of the "responsibility." It made what was already a powerful partnership much more of a real partnership. It was truly both of us. We grew comfortable with what lay ahead, and we grew comfortable discussing it.

Early on, I had thought about changing the direction of my brand-new lab to go after albinism. I think it is a common reaction among scientist parents. But I quickly convinced myself that I knew nothing about the eye; moreover, as I learned more about the pathophysiology, I convinced myself that there was little that could be done after the point that the retina had not developed normally. Furthermore, there was no easy path to editing Ruthie's genome at that time. I did imagine that genetic engineering could someday help kids who were diagnosed right after birth. It always seemed that the eye would be a suitable place to experiment with genetic engineering because of its accessibility. But I focused instead on just loving and supporting the child I had and not the one I wished I had.

Over the next few years, our initial fears and regrets morphed slowly into acceptance and then a full embrace of Ruthie with albinism. Sure, we had concerns. Top of the list was a worry that she would always be different. She would always be an outsider. These concerns were based on both the physical (she might not be able to play sports with her friends and she would look different) and the emotional (she might not have friends, she might not feel like she belongs, and she might be teased). Ultimately, we realized how lucky she was—and how lucky we were. All parents think their kids are special, but we knew this one was particularly special. Sure, she would run into windows, and she had enough bruises and bumps to warrant a call to child protective services. It was annoying for her and for us to remember her hat and sunglasses, and sunscreen application was always a challenge. But these were exceptions. Overall, life with Ruthie was joy. It was pure joy.

We began to learn about how the world saw differences. We started to think about how we saw differences. The local science museum had obtained an alligator they named "Claude the Albino Alligator." I remember when Ruthie's preschool planned a field trip there and Palmer was sure to remind the teachers that this was a potentially awkward moment. It went fine. Later, the museum ran a series of radio ads talking about why Claude lost his color. They were insensitive and in poor taste, and of course we never would have noticed if we did not have a child with albinism. Palmer convinced me that we needed to do something, so I wrote a letter to the UCSF Chancellor at that time, Sue Desmond-Hellmann, who was on the board of the museum, asking her to ask them to take the ad down. They did.

As Ruthie grew older, chaperoning field trips became one of my favorite activities, and it was not for the obvious reasons. You see, on field trips, I could watch what would happen when Ruthie moved to the front to see a demonstration or a piece of art. Without fail, this would elicit a chorus of protests from the students who naively assumed Ruthie was just being pushy. And like clockwork, her classmates would immediately scold the other children in what became a regular act of spontaneous and beautiful advocacy.

The theme here is that despite deriving from a stable germline mutation, Ruthie's albinism had evolved from a disability we all wished she never had to something we learned to accept to something we celebrated to something we cherished. The practical reality is that we briefly considered having a third child. This led to a series of what if conversations over whether we would want to have another kid with albinism. Would it be helpful for Ruthie to have a sibling to share this with her? Would it normalize her even more? Or was it unfair to that child, just as being legally blind was unfair to Ruthie?

In the end, we decided we would not have another kid, but we had decided that were we to change our minds, we would not do preimplantation genetic diagnosis, something that was readily available at that time. We would roll the genetic dice.

In the lab, I spend most of my time working with genetically modified mice. I knew that modifying the genome of a human would be possible someday. I knew that modifying human germ cells would be possible. But I did not really think of using genetic engineering to treat or prevent albinism until November 2015, when I read a Tweet from a scientist I did not know but respected: "Prediction: my grandchildren will be embryo-screened, germline-edited. Won't 'change what it means to be human.' It'll be like vaccination."

I read it a few times and then decided to respond: "Brings up so many hard questions. My daughter has *OCA2*. We did not have more kids but would not have screened. . . . But when she asks if she can see like the rest of us and just wants to read a book, I wonder if we are stupid."

My Tweet set off a long series of conversations with my friends, my family, my colleagues, the world, and importantly with Ruthie. These conversations continue today and will for the rest of time. While details change, the thread remains the same. We are blessed to have Ruthie in our lives and especially blessed to have Ruthie with albinism. Ruthie is immensely proud of who she is. At least for now, she feels strongly that she is who she is because of her albinism and, perhaps, despite it. While she still complains about her hat, sunglasses, and sunscreen, and while it is obvious she thinks hard while watching her older sister learn to drive, she remains steadfast in her commitment to loving herself and seeing herself and her albinism as one. It is clear to us that Ruthie is special and that not all kids with genetic disorders will ski, surf, or play on an AAU basketball team. Nor will all kids have the rich life she does, supported by an amazing community of friends and family. Some kids will not walk. Some kids will die. We absolutely believe there will be a role for gene editing for certain conditions that significantly affect morbidity or mortality. I can imagine that many targets for gene-editing-based therapeutics will be for obvious conditions that cause pain or suffering to children or conditions that significantly affect their families. There is exciting promise here, and the positions we have taken on albinism and Ruthie's treatment are strongly driven by our family's philosophy. Most importantly, we are 100% committed to listening to Ruthie; if she decides she wants to intervene in her condition, we support her.

In the end, what we have learned at this point in our journey above any other is that one cannot know what it is like to parent a child with a disability until one knows. Without knowing, it is practically impossible to make an informed decision about whether and how to intervene.

Update Fall 2023

Ruthie is now 17 and in high school. She is thriving academically, socially, and athletically. She continues to play basketball at a high level on her high school varsity team. In the spring of 2023, she also took up jumping events in track and field. Track is a sport we always thought might be good for her because low vision is not an obvious barrier to success. She was quite a natural, going on to set several school records in the first few weeks. It is not clear how track and field will fit into her future life, but it does seem as if she has talent and enthusiasm and it is a sport where she can compete on an even playing field (metaphorically). While she has never complained about the impact of her vision on her basketball abilities, I think she understands how difficult it is for her to play at a high level. It is an interesting thing to watch in a kid who says she is not daunted by her challenges but has clearly gravitated to a sport where her vision will have less or no impact. It is too early to tell if she will pursue this longer term, but in the interim, she is excited and committed. She loves her teammates and coaches and loves pushing herself in ways that I could never imagine.

On the gene-editing front, there has been a recent publication in the *New England Journal of Medicine* on the use of in vivo CRISPR-*Cas9* editing for ATTR amyloidosis (Gillmore et al. 2021). This development is exciting and is likely the beginning of many programs aimed at treating genetic diseases in vivo. For the time being, it appears that most of these programs will focus on the hepatocyte, given the apparent efficacy of editing these cells and other characteristics that make it favorable. What might this mean for albinism? Editing a whole embryo remains off the table for now, but there does seem to be a robust appetite for attempting to treat the eye. While albinism is a systemic condition, the most significant impact is on the eye and on visual acuity. It remains to be seen whether this approach will work in adults with albinism. The effect on the retina is developmental or, at least, very early postnatal. There are so many questions such as: Can you repair a highly diseased retina that has been damaged since birth? Is it only about editing the mutated gene? Will this lead to meaningful and significant improvements in visual acuity? How durable will it be? What are the short-, medium-, and long-term safety concerns? Indeed, there has been recent news of CRISPR successfully improving vision in patients with Leber congenital amaurosis (Stein 2021). This condition leaves patients with progressive and severe vision loss culminating in blindness in the first decade of life. So, the risk-reward is certainly different than in a less severe and nonprogressive condition such as albinism. For now, this is a fascinating space to watch, but Ruthie remains happy as she is and is currently not interested in any such future treatments. We have always said to each other and to her that we will all keep an open mind and that a safe and effective therapy to improve her vision might be something to consider down the line.

One thing that Ruthie does seem most bothered by right now is her lazy eye. Her amblyopia has bothered her more as she has aged. On a recent visit to her ophthalmologist, she asked about repairs. While there are certainly ways to fix the alignment, they would likely result in some amount of double vision. This is the old "you can fix the eye, but you cannot fix the brain" problem. I mention this only as a caution to remind us that even if we can eventually use gene editing or other approaches to "repair" the retina, we still need to overcome the obstacle of getting the brain to properly process all this new information. It is a recurring issue in the eye and is one more thing to keep in mind as we proceed.

When to Intervene

This conversation has great weight in our family in that we know, beyond a doubt, that had we been aware of Ruthie's condition before she was born, she would not be here today. She would have been filtered out as an embryo, or she would have been terminated. In the future, she might have been edited—perfected—fixed.

This is the crux of the conversation about the role of technology in medicine. Before our life with Ruthie, I did not consider strongly enough the distinction between choosing to intervene because we can versus because we should. That is it. We have at our fingertips these incredibly powerful tools that will permit us to do things that were unimaginable even 20 years ago. We can now diagnose genetic conditions prepregnancy. We can select genetically "optimal" embryos. Soon, we will be able to fix many of the broken ones.

But having Ruthie in our lives offers us a perspective that we never could have had before she was here. We passionately believe that Ruthie's presence in this world makes it a better, kinder, more considerate, more patient, and more humane place. It is not hard, then, to see that these new technologies bring the risk that the world will be less kind, compassionate, and patient if there are no more children like Ruthie. And, of course, the kids who inevitably end up here will be even less "normal" than they are today. The rest of the world will have an even harder time understanding and appreciating what they bring. What is most disturbing to me is that I know that even as a highly educated person in this field, I would have chosen to use the technology to remove or fix Ruthie if I had had the chance. Maybe it is a function of our training as doctors and scientists to strive to fix broken things, or maybe it is a function of our culture of perfection. I cannot say. What I can say is that I realize that life is one gigantic series of random, stochastic events and that the lottery that led to our existence today involved chance, the magnitude of which we can barely conceive. We consider ourselves to have won this game. We are better as parents, better as people, and better as a family to have had this experience of learning from Ruthie. We believe the world is a better place for having kids like her in it, and we want the world to think hard about whether it really wants to go down a path of engineering a world where there are no Ruthies.

Acknowledgments

I want to acknowledge my wife Palmer for being a spectacular partner; I am fortunate and grateful to have had this journey with her. I also want to acknowledge my daughter Mina; our dog McDuff; the rest of our family and friends; and Ruthie's friends, doctors, teachers, coaches, ski guides, and all the other people who have helped her thrive in this world. Lastly, of course, I want to acknowledge Ruthie herself.

REFERENCES

Gillmore, Julian D., Ed Gane, Jorg Taubel, Justin Kao, Marianna Fontana, Michael L. Maitland, Jessica Seitzer, et al. 2021 "CRISPR-Cas9 In Vivo Gene Editing for Transthyretin Amyloidosis." *N Engl J Med* 385 (6): 493–502. https://doi.org/10.1056/NEJMoa2107454.

Stein, Rob. 2021. "A Gene-Editing Experiment Let These Patients with Vision Loss See Color Again." *All Things Considered*. Aired September 29, 2021, on NPR.

8

Curing Cystic Fibrosis?

Sandra Sufian

When I read an article on August 9, 2021 in *The Telegraph,* titled "Cystic Fibrosis Cure on Horizon after Scientists Fix Genetic Mutation" (Knapton 2021), I immediately texted my doctor and asked: "Is this real?" As a historian of medicine and disability, who keeps up with the literature around CRISPR and cystic fibrosis (CF), I was immediately suspicious of such a headline because the news seemed to come out of nowhere, and the narrative is triumphalist and moralistic. The author's use of "cure on horizon" and "fix" reflects the pervasive cultural attitude that eliminating genetic disease is an incontrovertible "good." My colleague, Rosemarie Garland-Thomson, and I analyze this issue in our *Scientific American* op-ed (Sufian and Garland-Thomson 2021).

Like many other newspaper headlines, this one overpromises. Although genetic editing for CF is probably inevitable, the "cure on [the] horizon" is not as immediate as the headline suggests. The article describes work by scientists at the Hubrecht Institute in the Netherlands who successfully exchanged the "defective" code found in CF with "healthy" non-CF DNA in the lab, enabling cells to function "normally" (Knapton 2021). The scientists grew miniature organs, known as *organoids*, from human intestinal stem cells of CF patients. When they inserted the non-CF DNA into the tissue, the cells became swollen as they would in people without CF, thus proving to the team that their method had opened the Cystic Fibrosis Transmembrane Conductor Receptor (CFTR; Schwank et al. 2013). To perform the insertion, they used a modified technique of CRISPR called *prime editing.* Scientists consider it safer than the CRISPR/*Cas9* mechanism because it inserts new genetic material without accidentally causing damage to other areas of the DNA. The method is, therefore, thought to prevent unforeseen problems. Whether that is the case in humans remains to be seen. More recently, researchers at Yale University have conducted a mice study that successfully showed that simultaneous, systemic gene editing of multiple organs is possible. So is intravenous delivery of editing therapy (Wexler 2022). Like the prime editing case, studies in humans of this therapy will have to confirm these findings, including how long the corrective effects last.

To be sure, it is one thing to be able to insert new DNA into human stem cells or correct CF mutations in mice and another to do these techniques in a living person. Researchers and ethicists have recognized the need to adapt gene editing techniques to make them safe in humans. That could be a quick or lengthy process. Whatever the wait, there are certainly people with CF who would partake in somatic gene editing whenever it becomes available and affordable. Many families and individuals write about their hopes for a cure. They would welcome the potential impact of CRISPR technology to alleviate CF symptoms and lung function decline, hospitalizations, hours of doctor's visits, dozens of pills and medical treatments, and extremely high medical bills (Jiménez 2022). Taking up the same cure discourse, the Cystic Fibrosis Foundation (CFF) is now investing 500 million dollars through 2025 in a research program called the Path to the Cure, which seeks to develop treatments that can address the underlying cause of the disease using gene editing.

While scientists, bioethicists, the media, and the CFF focus on the prospect of using gene editing technology to cure CF, that possibility raises several nonscientific questions for someone like me, a middle-aged woman with CF who has lived through several transformations in, and reaped the benefits of, the clinical management of the disease. Those questions primarily deal with medical technologies' interaction with sociality.

One set of questions concerns the overdetermined, reductionist tendency that accompanies the use of diagnostic categories like CF in the quest for medical cures: What precisely do scientists, journalists, and the public imagine about CF when they think about eliminating it? Do they imagine the broad spectrum of disease or the countless times when we are not suffering? In most cases, the often naturalized, out-of-date image of CF often butts up against the rich, diverse experiences of actual people with the disease.

Another question about gene editing concerns the relationship between medical technology, illness and disability, cure, and subjectivity (Clare 2017). Part of living with a genetic disease like CF is the inability to delink my condition from myself; CF is an essential part of my being, not an appendage that one can splice away. Moreover, my identity, as well as many others', is closely tied to living within a patient and scientific community. Diagnostic categories alone cannot fully capture those associations. These perspectives adhere to Paul Rabinow's use of *biosociality*: the entangled cocreation of biological and social processes because of genetic technologies and knowledge (Fritsch 2016; Macgregor 2012). Rabinow argues that social groups form around genetic identity during this process. In addition, these groups develop traditions, narratives, and practices that create a cultural community (Mueller 2022b; Rabinow 1996). The CF community fits Rabinow's description of genetic biosociality closely, with extensive online activities, conferences, rituals, and regularized infection control norms and practices.

As the scientists at Hubrecht, Yale, and elsewhere work to cure CF in the coming years, this biosocial community is currently focusing on, and experiencing, the effects of another therapy. CFTR modulator therapy is a pill regimen that improves lung function so significantly that it drastically lengthens life expectancy and even brings some patients off the transplant list. Ironically, the company that developed CFTR modulator therapy, Vertex Pharmaceuticals, partnered with Moderna to undertake gene editing for CF (Businesswire 2020). Not every patient can take these medications owing to their genetic mutations. Still, the uptake and outcomes for those on modulators have reshaped patients' actual life prospects and horizons—our "disability futures" (Kafer 2013).

The prospect of a cure brought by CRISPR technology and the treatment that modulator therapy has provided share differences and similarities. The former seems more extensive, terminal, and unknown, whereas the latter is dramatic, still carries uncertainties, and comes close to, but does not reach, the finality of what we think of as a "cure." Whether a CRISPR cure is definitive is also uncertain; given that the technology alters genetic material and virtually all genetic expressions involve cascades, there will probably be unintended consequences with such manipulation (Sander Gilman, email communication with author, August 11, 2021). Now, we cannot know if a CRISPR cure will be seamless, if it will require repeated gene editing treatments, or if it will lead to other diseases or impairments. If so, we may need to revise our notion of "cure" as a single intervention that fixes everything when CRISPR is used on human bodies.

For its part, knowledge of the full extent of the physical, psychological, social, and cultural repercussions of modulator therapy on individual patients is still emerging. Medical practitioners and patients in the CF community are vigorously discussing and debating these implications. Modulator therapy, for instance, has improved lung function, lessened treatment burden, enabled pregnancy more easily and caused otherwise low-weight patients to gain weight (Taylor-Cousar 2020). People are returning to work and school, starting families, charting new paths, and expanding their prognostic imaginations (Mueller 2022a). But like every medicine, there are adverse iatrogenic effects: some people see their liver enzymes rise and experience unbearable mental health effects, memory loss, or rheumatic issues, to name a few. Obesity is now a problem in the CF community because of modulators. Individuals are recalibrating their body image while providers are raising concerns about heart disease with their adult patients, a topic rarely discussed only a few years ago. Despite modulators' significant benefits, some individuals cannot tolerate the side effects and have had to discontinue their use.

Modulator therapy is recasting our patient community along new lines. Gene editing will do the same. Whereas a CRISPR cure may eventually cause the CF community to disintegrate, the use and impacts of modulator therapy are creating new, more granular bioaffiliations within the CF community—ones organized around genetic mutation, specific side effects, and life stages and age, among others.

Here, the difference between gene editing and modulator therapy is stark: the unstated intention of gene editing is to make bioaffiliations entirely unnecessary. Suppose prime editing, for example, becomes feasible for and successful on humans. In that case, it is possible to imagine emerging communities structured around the moment a person swaps their DNA and becomes a person without CF (Rabinow 1996). Smaller communities may form based on lack of access to CRISPR or the extent of preexisting disease (previously acquired impairments or tissue damage) that CRISPR cannot repair. People with CF who are adults and undergo DNA replacement, for example, may strengthen their affiliations with others based on lingering medical, identity, social, and psychological issues related to chronic disease. In my conversations with people with CF about these possibilities, one woman stated that she thought that we would still "need each other even after CRISPR is used [on us], to work through *not* having the disease" (M. S., email communication with author, August 11, 2021). In that case, biosocial communities based on the specter or absence of disease could form, with support centered around the struggles of coping with new realities. Indeed, a second person confirms her belief in this prospect. She proffered that the alleged quick fix offered by CRISPR may create complicated work, school, social, family, and coping problems for people already living with CF.

The issue of how persons with CF relate to modulator therapy can be constructive for pondering how individuals will move forward with gene editing. Individuals with CF are weighing questions right now regarding modulator therapy and its drastic effects. On community group discussion boards on social media, people express their simultaneous elation, bewilderment, and struggles at redefining their future, saving for retirement, and constructing new identities (CF Facebook Group 2022; Mueller 2022a).

Individuals with CF, providers, and researchers are exploring questions like: How do people who have lived with a disease for decades suddenly cope with huge improvements that enable them to envision new lives? What happens to a person who has not worked because of CF and therefore has no work history but now can suddenly work? How does that person find a job? What might she be qualified for? (G. B., email communication with author, August 11, 2021).

In the case of gene editing, we can ask similar questions: How does an individual shed her identity as a person with CF postcure, or does she? How might the transition from having a disease to not materialize for different people? How would adults with CF-influenced forms of femininity and masculinity transform when the risk of mortality is no longer imminent? Just as manipulating genes in vivo would likely lead to a cascade of physical effects, so too would it involve a spiral of profound biosocial repercussions.

Further, the trajectories and questions for babies born with the genetics of CF who undergo a CRISPR cure will be entirely different from adults. Even now, modulator therapy has created a generational divide within the CF community. But in the case of CRISPR, will parents tell their children that they opted for genetic editing? Will those children later create ties according to a "cured" status or according to the possible sequelae of a cure? Will these sequelae cause more limitations in daily life than having CF (not to mention who would decide that)? What about children and adolescents living with CF who undergo a CRISPR cure (or will it only be approved for babies)? What might the permutations of such life changes look like for them? An even larger question might be: Should CRISPR be used to eliminate CF through germline editing so the disease is edited out of existence?

These kinds of questions are absent from sensational narratives like *The Telegraph* article. Even ethical debates about the use of CRISPR and the quest for cures tend to overlook questions of biosociality. Perhaps it is because they complicate the normalized idea of the strict, moral good of cure. But biosocial questions and those about the intersections of illness, disability, medical technology, and cure are critically important to consider because they insist upon examining the fears and joys of real people's lives with CF. They can acknowledge the desire of many CF patients to be done with the hours of treatments, expensive medicines, fatigue, time-consuming clinic visits and hospital stays, rejection and discrimination, and fears of death while also leaving space for thinking about the joys within biosocial communities and the possibility of losing them with gene editing. These include the joys of friendships; inside jokes; mutual, unspoken understanding; alliances with medical practitioners; experiential expertise about chronic disease, patienthood, and the body; and patient advocacy and activism. Biosocial questions allow us to consider that losing the culture of the CF community might cause significant grief for those who have grown up living with CF.

I feel lucky that I can take CFTR modulator therapy. I am undecided as to whether I would partake in a CRISPR "cure." What I do know, at the time of this writing, is that the prospect of using CRISPR to eradicate cystic fibrosis evokes a set of hypothetical quandaries that capture the heart of what it means to have a disability and, therefore, what it means to be human: to have hopes and plans about one's future while still being worried about its unfolding; to form, live, and thrive in community and to sustain supportive social relationships; to be open to and appreciative of life's opportunities but also to acknowledge one's limitations; to recognize one's deep knowledge about the body as it is lived (and not merely imagined or projected); and to acknowledge that people's bodies and minds vary while not marking some as "good" or "bad." Many people with chronic diseases understand that medical struggles come with wisdom, adaptability, acumen, and humility.

Though on its face, somatic gene editing may seem unproblematic—particularly compared to germline uses of this technology—the use of CRISPR will still likely raise complex, real-world dilemmas for individuals living with a genetic disease, for their biosocial affiliations, and for the contours of their life horizons. Though curing CF is a long-held goal of scientists, physicians, parents, foundations, and many patients, a cure may eventually threaten the survival of the larger CF community many of us hold dear.

REFERENCES

Businesswire. 2020. "Moderna and Vertex Establish New Collaboration to Treat Cystic Fibrosis Using Gene Editing." https://www.businesswire.com/news/home /20200916005915/en/.

(CF) Cystic Fibrosis Facebook Group. 2022. Series of Facebook posts.

Clare, Eli. 2017. *Brilliant Imperfection: Grappling with Cure*. Duke University Press.

Fritsch, Kelly. 2016. "Blood Functions: Disability, Biosociality, and Facts of the Body." *J Lit Cult Disabil Stud*, 10 (3): 347.

Jiménez, Marco. 2022. "One Day We Will Find a Cure for Cystic Fibrosis." *CF News Today*, May 18, 2022. https://cysticfibrosisnewstoday.com/31-days-of-cf/31-days-of-cf -one-day-find-cure/.

Kafer, Alison. 2013. *Feminist, Queer, Crip*. Bloomington: Indiana University Press.

Knapton, Sarah. 2021. "Cystic Fibrosis Cure on Horizon after Scientists Fix Genetic Mutation." *Telegraph*, August 9, 2021. https://www.telegraph.co.uk/news/2021/08/09/cystic -fibrosis-cure-horizon-scientists-correct-genetic-mutation/.

Macgregor, Casimir. 2012. "Genetic Biosociality in an Age of Biopower." *Sci Cult* 21 (4): 592–600. https://doi.org10.1080/09505431.2012.702747.

Mueller, Rebecca. 2022a. "Prognostic Imagination: Genetic Counseling amidst Therapeutic Innovation and Evolving Futures." J Genet Couns 32 (4): 762–67. https://doi .org/10.1002/jgc4.1660.

Mueller, Rebecca. 2022b. "The Genome and the Biome: Cystic Fibrosis at Six Feet Apart." PhD diss. University of Pennsylvania.

Rabinow, Paul. 1996. *Essays on the Anthropology of Reason*. Princeton: Princeton University Press.

Schwank, Gerald, Bon-Kyoung Koo, Valentina Sasselli, Johanna F. Dekkers, Inha Heo, Turan Demircan, Nobuo Sasaki, et al. 2013. "Functional Repair CFTR by CRISPR/ Cas9 in Intestinal Stem Cell Organoids of Cystic Fibrosis Patients." *Cell Stem Cell* 13 (6): 653–58. https://doi.org/10.1016/j.stem.2013.11.002.

Sufian, Sandy, and Rosemarie Garland-Thomson. 2021. "The Dark Side of CRISPR." *Scientific American*, February 16, 2021. https://www.scientificamerican.com/article/the -dark-side-of-crispr/.

Taylor-Cousar, Jennifer L. 2020. "CFTR Modulators: Impact on Fertility, Pregnancy, and Lactation in Women with Cystic Fibrosis." *J Clin Med* 9 (9): 2706. https://doi .org/10.3390/jcm9092706.

Wexler, Marisa. 2022. "Gene Editing Corrects CF Mutation in Multiple Organs All at Once." *Alls CF News Today*, October 7, 2022. https://cysticfibrosisnewstoday.com /news/gene-editing-corrects-cf-mutation-multiple-organs-mice/.

IV **DIVERSE VOICES**

9

CRISPR and Gene Editing

Why Indigenous Peoples and Why Now?

Krystal Tsosie

Note: My views as an Indigenous person reflect my own experiences.

An oft-stated trope tells us that genetic technology is outpacing our conversations related to ethics. It is also incorrect.

For decades, nay, for well over a century, scientists and philosophers in Western academic institutions have debated the ethical limits of genetic modification and its larger implications for humans—including whether it should even be done—long before current gene-editing technologies came into existence. The questions surrounding gene editing are not new. In fact, in terms of the potential for manipulating biological material, the proverbial writing has been on the wall for quite some time, especially as we look back on the ramp-up to current gene-editing tools.

CRISPR-*Cas*9 systems were predated by over a half-century of precursor biotechnical innovations, starting with early splicing experiments of the 1960s and 1970s, leading to the creation of recombinant DNA after the discovery of the site-selective cleavage power of restriction enzymes. Other genome-editing tools, such as the less effective zinc-finger nucleases (1985) and transcription activator-like effector nucleases (TALENs; 2009), were also developed before CRISPR. These targeted systems, in turn, were advancements on nontargeted means of gene editing such as irradiation and reliance on biologically innate gene-repair mechanisms of the pre-1960s. Hence, the trajectory of innovation related to gene editing has always been self-evident.

At a certain point, however, we must ask why researchers have not been focusing on ethics during the entirety of this technological evolution commensurate with the pursuit of advancing innovation. Further, if researchers in Western science failed to adequately contemplate these ethical questions with other academics across disciplines, then what engenders current inclusion of Indigenous peoples—who have been historically and continually disenfranchised in genomics as research subjects—in these discussions now?

Indigenous Peoples as Geneticists

Science is knowledge acquired through repeated, experiential observations. It is thus interesting that science vis-à-vis Western epistemologies is often privileged as being objective pursuits of truth, whereas Indigenous approaches to understanding are often relegated as alternative or subordinate knowledge systems (Tsosie and Claw 2020). This hierarchical dichotomy risks devaluing Indigenous traditional knowledges that have endured over time commensurate with or even exceeding the span of academic institutions. To exemplify, many food crops today are the result of selective breeding and agronomist techniques cultivated by Indigenous peoples worldwide across millennia. Embedded in many Indigenous peoples' cultures are nuanced understandings of genetically transmitted phenotypic traits across complex clan systems and societal-lineal kinship structures (Begay et al. 2020). It is important, therefore, to realize that Indigenous peoples have always been scientists and, in many ways, have served as geneticists and genetic pioneers.

The narrative of Indigenous peoples being anti-science is thus a reductive condescension that erroneously conflates Indigenous peoples' hesitance in serving as research subjects in genetic research with overall hesitance toward genetics (Lee 2015). There are sustained reasons, notwithstanding historical bioethical harms perpetrated on Indigenous communities by biomedical researchers, that have deterred many from participating in genomics research. Despite over a decade of efforts to increase inclusion of people from non-European genetic ancestry in research, fewer than one percent of research participants in genome-wide studies identify as Indigenous (Popejoy and Fullerton 2019). For many Indigenous peoples, concerns about gene editing are still yet on the horizon. Uncertainty about the uses of gene editing and usurpation of decision-making agency over *taonga* (Māori for "treasured") species integral to Indigenous ways of living are the more proximal concerns, especially when there does not appear to be clear and direct benefit to communities and the environment (Hudson et al. 2019). These concerns would, of course, be compounded for human applications.

To minimize Indigenous people's concerns or willful disengagement in gene editing as being anti-science is also to overly privilege the same agents that have always unilaterally benefited from the involvement of disenfranchised and disempowered peoples in research. Coupling the belittlement of Indigenous science as unobjective with pointing to Western or dominant approaches as objective is to commit fallacy. Subjectivity is introduced when only certain research questions or technological approaches are favored. Indeed, when we chase the flashiest genetic tool or favor only non-Indigenous interpretations of data, we also risk missing key information to inform our narratives. Unpacking this also entails transparently disclosing the power dynamics that benefit from the inclusion and extraction of Indigenous genomic data in this latest push for innovation. Without this critical lens, academics and industry partners risk perpetuating the same genetic harms onto Indigenous peoples, just with new and updated gene-editing technologies.

New Genetic Tools, Same Extraction?

It is important to realize that recent calls to increase inclusion of diverse populations in genetics are not new. A glance at the failure of genetic innovation to impart direct benefit to Indigenous peoples, whose genetic data and knowledges have been usurped for the furtherance of academia and industry, should give critical pause as to whether gene-editing tools will truly democratize genomics (Tsosie et al. 2021). (Any system that benefits most will still disenfranchise minoritized groups. Hence, we should be advocating for equity as opposed to democratization; the two terms are not synonymous.) DNA biomarkers from Indigenous communities outside of the United States, for instance, were collected by large-scale diversity projects of the 1990s and early 2000s through the Human Genome Diversity Project and the 1000 Genomes Project. While Indigenous peoples have been presented with vague promises that their participation in genomics research would, someday, lead to health interventions that would benefit them, to date Indigenous peoples are waiting for those promises to come to fruition. Meanwhile, for-profit companies such as those selling direct-to-consumer genetic ancestry tests tied to partnerships with pharmaceutical companies searching for the next blockbuster drug are increasingly looking to these "open access" databases for insight into Indigenous genomes (Tsosie 2022). Hence, we must question whether gene-editing tools will truly benefit Indigenous communities or whether they will continue to substantiate existing power imbalances at the expense of Indigenous peoples providing DNA.

For true transparency, the commercialization and co-optation of Indigenous genomes must also be examined within the constructs of social choice as well as power dynamics. Empowered agents design the technology under the influence of politics and market dynamics (Peña 2020). These decisions entail a level of social choice that is not always equitably presented to Indigenous peoples, whose genomic information is often collected without their full consent or knowledge, for instance by consumer-as-participant ancestry companies.

Unfortunately, Indigenous peoples cannot remain disengaged from genomics for long. If the field of genomics continues to advance, to include gene-editing tools, with unfettered access to Indigenous genomic data without full informed consent of the Indigenous peoples and without respect to Indigenous data sovereignty, then Indigenous genomes will continue to be usurped from Indigenous peoples. Even though federally recognized Tribal nations in the United States may choose to opt-out of genomics by establishing genetics moratoria, DNA from unprotected Indigenous groups will continue to be misappropriated. "There is no Indigenizing capitalism" (Kam'ayaam/Chachim'multhnii 2014).

Genetic Reductionism

While some Indigenous perspectives have relinquished the possibility that genetic modification would be of interest should it result in direct benefit to Indigenous communities, not much has been stated about the potential benefit of human germline modification. As mentioned, there is a hesitance in embracing genomics tools of any sort in Indigenous contexts. Notwithstanding the extraordinary timeline gap between the introduction of gene-editing tools and the time to actual benefit to Indigenous communities, this timeline disparity is even greater for heritable genome editing (HGE). But absent a collective Indigenous response for or against HGE, we can call into question genetic narratives that have already mischaracterized Indigenous peoples.

Of concern related to HGE are off-target effects, or unintended effects to untargeted parts of the genome that generate inadvertent consequences. While many state that CRISPR systems can mitigate these off-target effects, of less understanding is *bystander editing*, where base-editors can introduce multiple mutations at once at the on-target site. In short, we do not know how perfectly (or imperfectly) the body's own cell-repair mechanisms will work or whether there will be other alterations to the protein's function.

Of additional concern is using germline editing as a solution for defining what constitutes a normal human being. These are value-laden judgments that raise the danger of eugenics. Will certain genes associated with Indigenous peoples be judged less desirable and be replaced or modified using CRISPR?

Our knowledge of what happens to changes in the genome is still nascent. Frankly, we just do not have an adequate statistical sample size of diverse Indigenous peoples in genetic research studies to amply investigate these questions. Even if we did have this resource readily available, would the (mostly non-Indigenous) scientists interpret the results appropriately within the context of Indigenous histories they do not understand? Or would they use the data to further substantiate existing and biological reductive narratives that reflect their implicit biases of Indigenous peoples and cultures? Geneticists use whitewashed terms like *population bottlenecks* and balk at using more accurate terms such as *genocide* and *colonialism* to describe Indigenous peoples' population movements. This points to the fact that non-Indigenous scientists are not ready to face the harsh truths of our real Indigenous histories. Why, then, should we entrust them to narrate what constitutes our "normal" biological state?

Here's the other thing to consider: there are countless "mutations" associated with human adaptations to different environments. Unlike the point mutation in the *HBB* beta-globin gene in sickle cell disorder, most identified variants do not result in biochemically characterizable phenotypes. For instance, *FADS1* variants are associated with smaller body size and are protective for Indigenous Inuit populations in Greenland (Fox, Rallapalli, and Komor 2020). But we would not know the true extent of human genetic and phenotypic variations because well over 81% of participants in large-scale genome-wide association tests are of European ancestry (Popejoy and Fullerton 2019). This lack of inclusion of non-European people can lead to inaccurate interpretations of genome sequence data and overly simplified explanations of biological traits.

The failure to identify the evolutionary forces and biochemical mechanisms that underlie conditions that disproportionately affect Indigenous people can lead to incorrect conclusions that we are "genetically predisposed" to disorders when, in fact, there are more complicating, colonial narratives at play. For example, over 230 publications look at genetic predisposition to alcoholism in US Indigenous people when there are far more complex sociocultural etiologies at play.

Ultimately, using CRISPR raises the important question of whom we are empowering to make these decisions. Because of the past—and current—

exploitation of Indigenous people in genomics, we must ensure that we identify profit-seeking entities and guard against approving biomedical patents that seek to own pieces of our DNA. We must legally empower Indigenous communities and peoples to make decisions about who controls their DNA and how it will be used. And we must make certain that we Indigenous peoples are not used as guinea pigs to test CRISPR technologies or to modify our genomes.

REFERENCES

Begay, Rene L., Nanibaa' A. Garrison, Franklin Sage, Mark Bauer, Ursula Knoki-Wilson, David H. Begay, Beverly Becenti-Pigman, and Katrina G. Claw. 2020. "Weaving the Strands of Life (Iiná Bitł'ool): History of Genetic Research Involving Navajo People." *Hum Biol* 91 (3): 189–208. https://doi.org/10.13110/humanbiology.91.3.04.

Fox, Keolu, Kartik Lakshmi Rallapalli, and Alexis C. Komor. 2020. "Rewriting Human History and Empowering Indigenous Communities with Genome Editing Tools." *Genes* 11 (1): 88. https://doi.org/10.3390/genes11010088.

Hudson, Maui, Aroha Te Pareake Mead, David Chagné, Nick Roskruge, Sandy Morrison, Phillip L. Wilcox, and Andrew C. Allan. 2019. "Indigenous Perspectives and Gene Editing in Aotearoa New Zealand." *Front Bioeng Biotechnol* 11 (7): 70. https://doi.org /10.3389/fbioe.2019.00070.

Kam'ayaam/Chachim'multhnii (Cliff Atleo, Jr.). 2014. "Red Skin, White Masks: A Review." *Decol: Indig Educ Soc* 3 (2): 187–94.

Lee, Tanya H. 2015. "Genetics is Your Friend: Natives Not Anti-Science, Just Anti-Exploitation." *Indian Country Today*, November 10, 2015. https://indiancountrytoday .com/archive/genetics-is-your-friend-natives-not-anti-science-just-anti-exploitation.

Peña, Devon. 2020. "Gene-Drives and Indigenous Seed and Food Sovereignty." Indigenous Perspectives on Gene Editing in Health and Agriculture. Presented at CRISPRCon, September 3, 2020. https://www.keystone.org/indigenous-perspectives-on-gene-editing/.

Popejoy, Alice B., and Stephanie M. Fullerton. 2019. "Genomics Is Failing on Diversity." *Nature* 538 (7624): 161–64. https://doi.org/10.1038/538161a.

Tsosie, Krystal S. 2022. "Inclusion without Equity: The Need to Empower Indigenous Genomic Data Sovereignty in Precision Health." In *Remapping Race in a Global Context*, edited by Ludovica Lorusso and Rasmus Grønfeldt Winther, 148–54. London: Routledge.

Tsosie, Krystal S., and Katrina G. Claw. 2020. "Indigenizing Science and Reasserting Indigeneity in Research." *Hum Biol* 91 (3): 137–40. https://doi.org/10.13110/humanbiology .91.3.02.

Tsosie, Krystal S., Joseph M. Yracheta, Jessica Kolopenuk, Rick W. A. Smith. 2021. "Indigenous Data Sovereignties and Data Sharing in Biological Anthropology." *Am J Phys Anthropol* 174 (2): 183–86. https://doi.org/10.1002/ajpa.24184.

10

Do Trans/Humanists Dream of Electric Tits?

CRISPR and Transgender Bioethics

Florence Ashley

"Splice me, pharma daddy!"

—Ada-Rhodes Short, PhD

Whether it is titillating or fraught to bring CRISPR to bear on transgender life depends on one's beliefs about the value of transness and whether it is biological in nature. For individuals who believe that transness is a genetic phenomenon or due to exposure to prenatal hormones, CRISPR may lie uncomfortably close to correcting so-called genetic errors with the promise of high-tech conversion practices. For those who, like me, believe that gender identity emerges from an intricate interlacing of disparate factors, none uniquely identifiable in making us who we are, our fears about CRISPR may fall by the wayside as we begin to dream of high-tech customized hormonal therapies.

The bioethical implications of CRISPR for trans people turn on the relative likelihood and (un)desirability of the two scenarios. I will consider each in turn before offering some reflections on the hazards of trans/humanist dreams. Rather than offering definitive bioethical conclusions, I approach the topic in an exploratory mood, mapping out issues and possibilities that should be kept in mind when thinking about the ethics of CRISPR in relation to trans communities.

CRISPR as High-Tech Conversion Practice

Conversion practitioners have long deployed medical knowledge and technologies to undermine trans and queer existence. By conversion practices, I am referring to a wide variety of sustained efforts to change, discourage, or repress people's gender identity, gender expression, or sexual orientation, including attempts to prevent people from being or growing up to be trans or queer (Ashley 2021, 2022). Licensed and unlicensed professionals have used electroshock therapies, lobotomies, and hormonal treatments to bring patients into the cisheteronormative fold—sometimes with their consent, sometimes not—all-too-often with disastrous results (Green et al. 2020; Madrigal-Borloz 2020). While trans conversion practices are opposed by countless reputable professional associations, recent years have seen a surge of scientific and lay ideologues gesturing toward or outright promoting a revival of such trans-hostile practices. Conjuring a moral panic, proponents of conversion practices are portraying youth as confused subjects of social contagion who falsely believe themselves trans as an easy solution to their trauma or internalized homophobia and misogyny (Ashley 2019, 2020). Although unfounded, these claims have enjoyed significant acceptance and promotion by conservative lawmakers (McGuire 2021).

There is little doubt that some people would enthusiastically use CRISPR on zygotes, germ cells, or somatic cells to prevent others from growing up trans or to change their gender identity once it develops. It is unclear whether existing legal bans on conversion practices would cover all such uses of CRISPR and, more gravely, conversion practices remain both legal and common throughout most of the world. The easy access to CRISPR only heightens the risk that it will be used to eradicate trans existence if possible. This concern brings me to two questions: is it likely to succeed and would that be bad? My answers are respectively no and yes.

Many scientists appear to believe that we will one day identify genetic or hormonal causes for being trans (Graves 2019). Etiological research is profitable, and numerous teams have secured funding to identify a "trans gene," corroborate the existence of a "trans brain," or hypothesize about the influence of prenatal hormones on gender identity (Austrian Science Fund grant P23021; German Science Foundation grant HA 3202/7-3; and NIH grants 1R01HD087712 and 5R01HD087712). Such studies are often conducted under the altruistic, albeit naïve, belief that proving the biological foundations of transness would lead to greater social acceptance and secure legal rights tied to immutability (Diamond and Rosky 2016; Schüklenk et al. 1997). Rehashing earlier debates about the "gay gene," many trans scholars and scientists have opposed these research programs as misguided, wasteful, and dangerous and criticized their binary and gender essentialist assumptions (Caselles 2018; Mulkey 2021; Schüklenk et al. 1997). For those committed to intragender diversity and the nonbinary nature of gender, these studies can be quite suffocating. Even under the best of lights, they carry all the trappings of protoeugenics.

Despite decades of research, strong candidates for the gay gene or trans gene have remained elusive. While such genetic studies attract significant attention, they can be prone to false positives when using the conventional significance threshold of $\alpha=0.05$, due to the sheer size of the human genome. As the replication crisis in many disciplines teaches us, statistically significant results can and do occur by chance. And even if we set aside the multiple comparisons problem in whole-genome and whole-exome research, reported genetic influences on gay or trans identity appear minuscule. From a practical standpoint, these genetic contributions are far from the necessary or sufficient conditions that would effectively power high-tech conversion practices. For example, a recent study on hormone-signaling genes found that trans women were far more likely to have a particular genotype of a specific gene ($p=0.009$), yet only 37.4% of the trans women had it compared to 26.6% of the cis men (Foreman et al. 2019). Polygenic models may fare better at predicting transness (Polderman et al. 2018), but their success remains highly speculative, and the risks known and unknown of using CRISPR sharply increase as more genes are targeted. Moreover, there is no guarantee that editing genes after substantial fetal development, birth, or infancy would succeed in altering or preventing gender identity, even if genetic causes could be narrowly isolated (Arnold and Chen 2009; Levine and Mullins 1964; McCarthy et al. 2009). If trans genes could be identified, their role in development could well be irreversible. Acting on prenatal hormones might be more feasible, but unlikely to be any more fruitful. Studies on prenatal hormones, often done by comparing finger lengths—perhaps my childhood bullies were onto something—fare little better (Sadr et al. 2020). In any case, the role CRISPR could play in relation to prenatal hormones is not evident—altering parental genes, perhaps?

While many people would certainly like to use CRISPR as a high-tech trans conversion practice—and some will likely do so regardless of effectiveness—its successful deployment toward that goal strikes me as implausible. If gender identity emerges from the irreducibly complex interactions of genetic, hormonal, environmental, social, and psychological factors, as I believe it does, then the most conversion practitioners can hope for is perhaps a few percent fewer trans folks—which could just as easily backfire given the intractable opacity of causal influences. But just in case of misuse, we should stop funding studies into the origins of transness. There are better uses for the money.

Regardless of plausibility, CRISPR raises important ethical questions about what makes trans conversion practices ignoble (Earp, Sandberg, and Savulescu 2014). Much advocacy against conversion practices has emphasized the psychological harm of attempting to prevent children from growing up trans or attempting to change their gender identity once developed. This harm is often narratively linked to biological accounts of gender identity, although not always. Just because a car's direction is not fixed does not mean you can safely make it change lanes by ramming into it. CRISPR, however, conjures the specter of psychologically harmless conversion practices. If gender identity is genetic and we can change genes, it might be possible to turn a trans person cisgender without any negative psychological consequences. The possibility is only theoretical. Genetic brain changes and alterations to someone's sense of self could very well cause physiological and/or psychological harm. But, assuming harmlessness, would it still be unethical?

I may be biased as a transfeminine person, but I would answer that yes, it would still be unethical. While conversion practices are agents of untold harm for those subjected to them, they also perpetuate dehumanizing and degrading ideologies toward trans communities writ large. Knowing the extent of others' hatred toward your very existence is a heavy load to carry (Verrelli et al. 2019). In a world that devalues trans lives, where trans people face horrendous stigma, harassment, discrimination, and violence, and where over a third of trans people attempt suicide (James et al. 2016), I cannot fathom tolerating practices that try to prevent people from being trans or seek to eliminate us from society. Is that not, after all, what lies at the heart of the immorality of eugenics? Not just harm—although harm indeed—but inequality and dehumanization. Harm, not just to the individual, but to the very moral fabric of society.

CRISPR as High-Tech Medical Transition

Would it not be great to eschew hormone therapy in favor of CRISPR-induced endogenous hormone production? Instead of my daily cocktail of estrogen and progesterone pills, a little bit of gene editing and my body would rev up the production of all the hormones my genetic makeup long denied me. The prospect is certainly appealing—maybe enough for me to ignore that little voice in my head whispering that we should not be so pressed to play

God, and have we not learned anything from Icarus (Ledford 2020) or Lexi (Peters 2017)?

Transition is never as easy nor as effective as we want it to be. For trans communities, the use of CRISPR on somatic cells beckons a vague and elusive dream of ease. Precisely because clinical uses of CRISPR seem so distant, its possibilities seem endless. Not only could it facilitate hormone-related care by replacing pills and injections with endogenous production, but it offers the promise of customization. In surgery, gene editing could potentially be used to prevent visible scarring, alter tissue type (for instance, changing skin into mucous membrane), and grow tissue and organs for xenotransplantation (Roh, Li, and Liao 2018). Personalized hormone regimens are still largely unknown in trans health care, with microdosing estrogen and testosterone still in their infancy outside of do-it-yourself spheres. While dosage can alter the effects of hormones, it is not yet possible to pick-and-choose results and some would like to change their bodies in ways that transcend what hormones allow. For many trans people, no available transition-related medical intervention lets them change what they want about their bodies (Galupo, Pulice-Farrow, and Pehl 2021; Vincent 2019). If you want your voice to drop on testosterone, you must accept clitoral and beard growth. But with CRISPR? Maybe not. For instance, clitoral and beard growth could potentially be inhibited by developing techniques that localize hormone uptake or production or, more likely, by acting on genes specific to pilosity or clitoral development.

Trans health has historically been stuck in a binary model that casts medical transition as a movement toward male and female bodies, denying nonbinary identities or rendering them liminal (Bradford and Syed 2019; Riggs et al. 2019; Vipond 2015). Customizability breaks free of bodily bimodality, throwing wide open the door to medical androgenization and the proliferation of gender possibilities. At the same time, customization has the potential to further undermine the cis–trans distinction as people understood to be cisgender pursue medical interventions long interwoven with our cultural understandings of transness. Would some cis women not want lower voices, more facial hair, or larger clitorises? Would some cis men not want softer skin, thinner body hair, different orgasms, or breast tissue? CRISPR also holds the potential of blurring the line between gender and trans-species technologies (Weaver 2014). With feline traits socially coded as feminine, would CRISPR create a possibility for whiskers? (This one is for the cat girls, cat boys, and all other cat friends.) And if we understand tattoos and piercings as opportunities to express our gender, why would bioluminescent breasts not also be? Do trans/humanists dream of electric tits?

Dreams of undoing gender may be little more than dreams. As Sara Cohen Shabot (2006) has pointed out, the cyborg of feminist dreams poses a risk of reinforcing the oppressive categories it seeks to destabilize. I have little doubt that doctors will pursue the perfection of binary gender categories long before they develop technologies that pursue androgyny. Current forms of transgender health care are deeply influenced by normative ideals of white femininity and masculinity that supersede the importance of individual bodily goals (Ashley and Ells 2018; Gill-Peterson 2018; Plemons 2019). In a patriarchal world, it is not surprising that reifying rigid gender categories takes precedence over undermining them. Terminator's T-800 (Arnold Schwarzenegger) and T-X (Kristanna Loken) are stark examples of how cyborg gendering can turn out when left to the whims of prevailing social power structures. CRISPR could end up reinforcing the idea that there is something wrong with "looking trans," blaming those who do not wish to blend in with cisnormative society for their own oppression. Dominant ideological systems are often the first to claim the benefits of biotechnological development (Bliss 2015; Mire 2020; Morgan 2011; Riska 2009; Schmitz 2021).

Behind all its speculative grandiosity, CRISPR's greatest appeal for trans lives may be one of distributive justice. In my nonexpert estimate, the most plausible use of CRISPR in trans health is in altering the body's endogenous hormone production. Pulling out my receipts, my estradiol and progesterone pills would cost me around $CAD 2,000 per year without insurance. Even with insurance, $CAD 400 per year can be prohibitive given the overwhelming poverty rates in trans communities (Arps et al. 2021). Hormone prescriptions also entail finding and keeping a trusted doctor, taking time off work for appointments, and subjecting yourself to systems of medical gatekeeping and discipline. Unsurprisingly, access to transition-related medical care is highly stratified along socioeconomic and racial lines (Gill-Peterson 2018; Scheim et al. 2017). Social and financial barriers to CRISPR may be far lesser, especially on the do-it-yourself scene (Rotondi et al. 2013), bringing hope of a more egalitarian trans health landscape.

Trans/Humanist Dreams, Trans/Humanist Hype

As I wrote this essay, I butted against my limited knowledge of CRISPR. That ignorance nourished my speculations, turning them into monumental daydreams of gender upheaval. It was all too easy to let myself be carried away. However appealing, these daydreams are not benign. They create hopes, expectations, and wishes that inform our attitudes toward the world. When writing about CRISPR, we bioethicists should attend to the ethical implications of hype (Caulfield 2016).

CRISPR is hyped. Its possibilities are widely disseminated among the lay population, often through sensationalistic articles that exaggerate the range of clinical possibilities and how close we are to achieving them. CRISPR's prospects are the flying cars of today. We do not know how CRISPR will be usable in the future and whether it will be possible to deploy it as a transition-related medical technology. Its successes might turn out to be far more modest than hoped, as much of our imagination marches on unbound by technoscientific credibility. For trans individuals, speculative nonfiction about gene editing can generate life-sustaining hopes and expectations or create disappointment. Too many trans people have taken their own lives, distraught that they would never reach their embodiment goals. In her suicide note, Leelah Alcorn, a teenager, heartbreakingly wrote: "I'm never going to transition successfully, even when I move out. I'm never going to be happy with the way I look or sound" (Lowder 2014). The hope of technological progress can foster life. It can give hope. It can serve as a future to work toward, energizing current demands for more accessible, personalized trans health care. Yet I have also known trans people who were suicidal because they had cultivated unrealistic expectations and came crashing down as they realized that the expectations were naught more than wishful thinking. Escapism, even without the danger of disappointment, can trap you in a mindset of waiting and prevent you from living your best present. Which way will CRISPR bring people as it strolls down the trans social imaginary?

The hype of CRISPR also impacts trans health clinicians and family members. CRISPR dreams of a normatively gendered future for one's patients or children can reinforce present obliviousness and resistance to the infinite diversity of trans embodiment goals and lives beyond the gender binary. CRISPR's hype may hinder progress in trans health, reinforcing the desire to make trans people and especially trans youth indistinguishable from cis people instead of moving us toward individualized care that centers on patients' desires and self-understanding. In a similar vein, CRISPR may reinvigorate conversion practitioners in their crusade against trans existence.

Predicting scientific futures is difficult. Time and time again, our predictions turn out to be erroneous. I have yet to drive a flying car. Given the underexplored consequences of hype on psychology and society, we should be more careful when speculating about the potential of CRISPR. Debates around the ethics of using CRISPR, too, can give an impression of immediacy that does not track how far along we are in developing human gene editing. With the clinical deployment of CRISPR still a considerable distance away, we bioethicists should begin by discussing the ethics of hype. Caught in the daydreams of CRISPR's endless possibilities, let us not forget to fight for a better world.

Acknowledgments

I would like to thank Professor Ada-Rhodes Short, PhD, for her feedback and for providing me with a most spiffy epigraph. I also owe thanks to Reubs J. Walsh, Eartha Mae Guthman, Sofia K. Forslund, Rebecca Schalkowski, Cal Horton, Em Rabelais, and Anna Horvath for their helpful comments.

REFERENCES

Arnold, Arthur P., and Xuqi Chen. 2009. "What Does the 'Four Core Genotypes' Mouse Model Tell Us about Sex Differences in the Brain and Other Tissues?" *Front Neuroendocrinol* 30 (1): 1–9. https://doi.org/10.1016/j.yfrne.2008.11.001.

Arps, Frédéric S. E., Sophia Ciavarella, Jelena Vermilion, Rebecca Hammond, Kelendria Nation, Siobhan Churchill, Meghan Smith, et al. 2021. "Report—Health and Well-Being among Trans and Non-Binary People Doing Sex Work." *Trans PULSE Canada* (blog). March 30, 2021. https://transpulsecanada.ca/results/report-health-and-well-being-among-trans-and-non-binary-people-doing-sex-work/.

Ashley, Florence. 2019. "Homophobia, Conversion Therapy, and Care Models for Trans Youth: Defending the Gender-Affirmative Model." *J LGBT Youth* 17 (4): 361–83. https://doi.org/10.1080/19361653.2019.1665610.

Ashley, Florence. 2020. "A Critical Commentary on 'Rapid-Onset Gender Dysphoria.'" *Sociol Rev* 68 (4): 779–99.

Ashley, Florence. 2021. "Reparative Therapy." In *The SAGE Encyclopedia of Trans Studies*, edited by Abbie E. Goldberg and Genny Beemyn, 2:713–17.

Ashley, Florence. 2022. *Banning Transgender Conversion Practices: A Legal and Policy Analysis*. Vancouver: UBC Press.

Ashley, Florence, and Carolyn Ells. 2018. "In Favor of Covering Ethically Important Cosmetic Surgeries: Facial Feminization Surgery for Transgender People." *Am J Bioeth* 18 (12): 23–25. https://doi.org/10.1080/15265161.2018.1531162.

Bliss, Catherine. 2015. "Biomedicalization and the New Science of Race." In *Reimagining (Bio)Medicalization, Pharmaceuticals and Genetics*, edited by Susan Bell and Anne Figert, 175–96. New York: Routledge. https://doi.org/10.4324/9781315760926.

Bradford, Nova J., and Moin Syed. 2019. "Transnormativity and Transgender Identity Development: A Master Narrative Approach." *Sex Roles* 81 (5–6): 306–25. https://doi.org/10.1007/s11199-018-0992-7.

Caselles, Eric Llaveria. 2018. "Dismantling the Transgender Brain." *Grad J Soc Sci* 14 (2): 135–59.

Caulfield, Timothy. 2016. "Ethics Hype?" *Hastings Cent Rep* 46 (5): 13–16. https://doi.org/10.1002/hast.612.

Diamond, Lisa M., and Clifford J. Rosky. 2016. "Scrutinizing Immutability: Research on Sexual Orientation and U.S. Legal Advocacy for Sexual Minorities." *J Sex Res* 53 (4–5): 363–91. https://doi.org/10.1080/00224499.2016.1139665.

Earp, Brian D., Anders Sandberg, and Julian Savulescu. 2014. "Brave New Love: The Threat of High-Tech 'Conversion' Therapy and the Bio-Oppression of Sexual Minorities." *AJOB Neurosci* 5 (1): 4–12. https://doi.org/10.1080/21507740.2013.863242.

Foreman, Madeleine, Lauren Hare, Kate York, Kara Balakrishnan, Francisco J. Sánchez, Fintan Harte, Jaco Erasmus, et al. 2019. "Genetic Link between Gender Dysphoria and Sex Hormone Signaling." *J Clin Endocrinol Metab* 104 (2): 390–96. https://doi.org/10.1210/jc.2018-01105.

Galupo, M. Paz, Lex Pulice-Farrow, and Emerson Pehl. 2021. "'There Is Nothing to Do about It': Nonbinary Individuals' Experience of Gender Dysphoria." *Transgend Health* 6 (2): 101–10. https://doi.org/10.1089/trgh.2020.0041.

Gill-Peterson, Jules. 2018. *Histories of the Transgender Child*. Minneapolis: University of Minnesota Press.

Graves, Jenny. 2019. "How Genes and Evolution Shape Gender—and Transgender—Identity." *Conversation*, January 23, 2019. https://theconversation.com/how-genes-and-evolution-shape-gender-and-transgender-identity-108911.

Green, Amy E., Myeshia Price-Feeney, Samuel H. Dorison, and Casey J. Pick. 2020. "Self-Reported Conversion Efforts and Suicidality among US LGBTQ Youths and Young Adults, 2018." *Am J Public Health* 110 (8): 1221–27. https://doi.org/10.2105/AJPH.2020.305701.

James, Sandy E., Jodie L. Herman, Mara Keisling, Lisa Mottet, and Ma'ayan Anafi. 2016. "The Report of the 2015 U.S. Transgender Survey." Washington, DC: National Center for Transgender Equality.

Ledford, Heidi. 2020. "CRISPR Gene Editing in Human Embryos Wreaks Chromosomal Mayhem." *Nature* 583 (7814): 17–18. https://doi.org/10.1038/d41586-020-01906-4.

Levine, S., and R. Mullins. 1964. "Estrogen Administered Neonatally Affects Adult Sexual Behavior in Male and Female Rats." *Science* 144 (3615): 185–87. https://doi.org/10.1126/science.144.3615.185.

Lowder, J. Bryan. 2014. "Listen to Leelah Alcorn's Final Words." *Slate*, December 31, 2014. https://slate.com/human-interest/2014/12/leelah-alcorn-transgender-teen-from-ohio-should-be-honored-in-death.html.

Madrigal-Borloz, Victor. 2020. *Practices of So-Called "Conversion Therapy."* A/HRC/44/53. Geneva: United Nations Human Rights Office of the High Commissioner.

McCarthy, Margaret M., Anthony P. Auger, Tracy L. Bale, Geert J. De Vries, Gregory A. Dunn, Nancy G. Forger, Elaine K. Murray, et al. 2009. "The Epigenetics of Sex Differences in the Brain." *J Neurosci* 29 (41): 12815–23. https://doi.org/10.1523/JNEUROSCI.3331-09.2009.

McGuire, Lillian. 2021. "Outlawing Trans Youth: State Legislatures and the Battle over Gender-Affirming Healthcare for Minors." *Harvard Law Rev* 134: 2163–85.

Mire, Amina. 2020. *Wellness in Whiteness: Biomedicalisation and the Promotion of Whiteness and Youth among Women*. New York: Routledge.

Morgan, Kathryn Pauly. 2011. "Foucault, Ugly Ducklings, and Technoswans: Analyzing Fat Hatred, Weight-Loss Surgery, and Compulsory Biomedicalized Aesthetics in America." *Int J Fem Approaches Bioeth* 4 (1): 188–220. https://doi.org/10.3138/ijfab.4.1.188.

Mulkey, Nat. 2021. "The Search for a 'Cause' of Transness Is Misguided." *Scientific American*, March 23, 2021. https://www.scientificamerican.com/article/the-search-for-a-lsquo -cause-rsquo-of-transness-is-misguided/.

Peters, Torrey. 2017. *Infect Your Friends and Loved Ones*. CreateSpace.

Plemons, Eric. 2019. "Gender, Ethnicity, and Transgender Embodiment: Interrogating Classification in Facial Feminization Surgery." *Body Soc* 25 (1): 3–28. https://doi.org /10.1177/1357034X18812942.

Polderman, Tinca J. C., Baudewijntje P. C. Kreukels, Michael S. Irwig, Lauren Beach, Yee-Ming Chan, Eske M. Derks, Isabel Esteva, et al. 2018. "The Biological Contributions to Gender Identity and Gender Diversity: Bringing Data to the Table." *Behav Genet* 48 (2): 95–108. https://doi.org/10.1007/s10519-018-9889-z.

Riggs, Damien W., Ruth Pearce, Carla A. Pfeffer, Sally Hines, Francis White, and Elisabetta Ruspini. 2019. "Transnormativity in the Psy Disciplines: Constructing Pathology in the Diagnostic and Statistical Manual of Mental Disorders and Standards of Care." *Am Psychol* 74 (8): 912–24. https://doi.org/10.1037/amp0000545.

Riska, Elianne. 2009. "Gender and Medicalization and Biomedicalization Theories." In *Biomedicalization*, edited by Adele E. Clarke, Laura Mamo, Jennifer Ruth Fosket, Jennifer R. Fishman, and Janet K. Shim, 147–70. Durham, NC: Duke University Press. https://doi.org/10.1215/9780822391258-005.

Roh, Danny S., Edward B.-H. Li, and Eric C. Liao. 2018. "CRISPR Craft: DNA Editing the Reconstructive Ladder." *Plast Reconstr Surg* 142 (5): 1355–64. https://doi.org/10.1097 /PRS.0000000000004863.

Rotondi, Nooshin Khobzi, Greta R. Bauer, Kyle Scanlon, Matthias Kaay, Robb Travers, and Anna Travers. 2013. "Nonprescribed Hormone Use and Self-Performed Surgeries: 'Do-It-Yourself' Transitions in Transgender Communities in Ontario, Canada." *Am J Public Health* 103 (10): 1830–36. https://doi.org/10.2105/AJPH.2013.301348.

Sadr, Mostafa, Behzad S. Khorashad, Ali Talaei, Nasrin Fazeli, and Johannes Hönekopp. 2020. "2D:4D Suggests a Role of Prenatal Testosterone in Gender Dysphoria." *Arch Sex Behav* 49 (2): 421–32. https://doi.org/10.1007/s10508-020-01630-0.

Scheim, Ayden I., Xuchen Zong, Rachel Giblon, and Greta R. Bauer. 2017. "Disparities in Access to Family Physicians among Transgender People in Ontario, Canada." *Int J Transgend* 18 (3): 343–52. https://doi.org/10.1080/15532739.2017.1323069.

Schmitz, Sigrid. 2021. "TechnoBrainBodies-in-Cultures: An Intersectional Case." *Front Sociol* 6: 1–16. https://doi.org/10.3389/fsoc.2021.651486.

Schüklenk, Udo, Edward Stein, Jacinta Kerin, and William Byne. 1997. "The Ethics of Genetic Research on Sexual Orientation." *Hastings Cent Rep* 27 (4): 6–13. https://doi .org/10.2307/3528773.

Shabot, Sara Cohen. 2006. "Grotesque Bodies: A Response to Disembodied Cyborgs." *J Gend Stud* 15 (3): 223–35. https://doi.org/10.1080/09589230600862026.

Verrelli, Stefano, Fiona A. White, Lauren J. Harvey, and Michael R. Pulciani. 2019. "Minority Stress, Social Support, and the Mental Health of Lesbian, Gay, and Bisexual Australians during the Australian Marriage Law Postal Survey." *Aust Psychol* 54 (4): 336–46. https://doi.org/10.1111/ap.12380.

Vincent, Ben. 2019. "Breaking Down Barriers and Binaries in Trans Healthcare: The Validation of Non-Binary People." *Int J Transgend* 20 (2–3): 132–37. https://doi.org/10.1080/15532739.2018.1534075.

Vipond, Evan. 2015. "Resisting Transnormativity: Challenging the Medicalization and Regulation of Trans Bodies." *Theory in Action* 8 (2): 21–44. https://doi.org/10.3798/tia.1937-0237.15008.

Weaver, Harlan. 2014. "Trans Species." *Transgend Stud Q* 1 (1–2): 253–54. https://doi.org/10.1215/23289252-2400100.

V   THE DILEMMA OF CONTROLLING THE FUTURE

11

Velvet Eugenics

In the Best Interests of Our Future Children?

Rosemarie Garland-Thomson

The Modern Existential Dilemma

We always get to this difficult conversation one way or another when I talk to friends who have kids with disabilities. It goes like this: "If there had been a test for autism when my wife was pregnant with our son," my close friend tells me, "she would definitely have had an abortion." He tells me this with candor because he knows I know that this does not mean that he regrets having his son, who is now grown up.

Parents of children with disabilities are usually rightfully wary to engage in such conversations, but perhaps because I have a significant congenital genetic disability myself, we talk together about this with intimacy and mutual understanding. With his reflection about what we now consider reproductive choice, he is reaching back to where he was in his life and toward the person he was before his son was born—what philosophers call our "then self." This act of self-consciousness, the reflexive human habit of reimagining then through the knowledge of now, causes a catch in his throat. "What if," I press on, "you had not had this son but instead, perhaps, another child without autism or no child at all?" He pauses, trying to imagine his life without his son or with another in his place. He gently touches my arm and slightly grimaces under the task of conjuring a response. "I would not be who I am now," he tells me.

Without my having to ask, he goes on to say how parenting a child with a significant disability transformed his own life and made his family what it is. My friend explains that having a child with the characteristics we think of as autism shaped not just the person he has become but also his relationships with everyone in his family, his understanding of the world, and the meaning of his life itself. His life, as he understands it at the moment of our conversation, benefited from having that boy and would have been diminished without him. The catch in my friend's throat during our conversation bears witness to the burden of knowledge and the fragility of our future-making endeavors. The idea that he would have selected against bringing that son into his life and the thought of life without him are now unimaginable.

How do we account for this contradiction between my friend's then self and his now self, between what he knew then and what he knows now? What is counterintuitive to many of us is that my friend suffers not because he has a son who has some of the characteristics we think of as autism. Rather, he knows how he might have acted in the past based on medical information that would have caused him to not have that beloved son now. In other words, what my friend's past self considered to be his best interests were, in fact, the opposite of what his now self recognizes as his best interests.

Many parents of children with disabilities, of course, may suffer in numerous ways because of the disabilities or the decisions they made about their children. Some parents, no doubt, suffer because they did not act to prevent the birth of a child or because they cannot rescue their child from suffering. An expansive narrative literature—from blogs to popular articles, memoirs, and academic books—documents experiences ranging from inspiration to torment about parenting children with conditions that cause them and their families to suffer. The moral instruction these stories offer us is how complex the network of suffering turns out to be in the lived experience of childhood illness. What is clear is that the medical narratives of a child's diagnosis, treatment, and prognosis play out very differently depending on each family's social, geographic, economic, and spiritual context. What seems most significant in these narratives is not the medical experience of disease but rather the ends of suffering itself that families take from the experience of shared childhood suffering.

The nature of a disability and its livability certainly affect whether and how everyone involved may suffer. Before I return to my friend's story about his son and the distinctively modern existential burden of needing to make decisions in the present that will yield intended but uncertain future outcomes, I turn to the historical sources that now give rise to his dilemma.

Eugenic Science

With eugenics, modern science expanded from describing the world to shaping it. Modern medicine went beyond relieving human suffering to preventing what it defined as the sources of human suffering. The distinctly modern enterprise of medical science sought to determine the future of our bodies through eugenics, the dominant scientific theory of the late nineteenth and early twentieth centuries and—arguably—beyond. *Eugenics*, meaning "good genes," offered a knowledge system secured by the truth value of scientific rather than divine authority.

Eugenics was both pseudoscience (drawing on racist agendas) and modern science (understanding certain diseases as genetically based) at the time. It was a set of rationales and actions with which modern nation states intentionally tried to shape and improve their citizenry through selective breeding. In this sense, eugenics is a systematic way to compare biological differences to determine which ones are better (Cohen 2016; Duster 1990; Fries 2017; Garland-Thomson 2007, 2017; Lombardo 2008; Paul 1995; Proctor 1998). Modern science thus annulled the family of man as all God's children and instituted the mechanics of human heredity to bind us to one another.

Endowed with the rising authority of science to explain and guide human existence, eugenic science aimed to prevent those the ascendant group considered alien from entering communities by inventing *health* as the gatekeeper that could produce a standard citizen compliant with the values of the dominant order. Controlling future population outcomes through present action is a form of human rationalization that the philosopher Ian Hacking (2006) has called "making up people."

The Ideology of Health

With the science of eugenics, modern medicine goes beyond the traditional aim of ameliorating human suffering and the immediate effects of mortality. The business of modern medicine, then, is to sort human variations into the opposing categories of disease and nondisease, or what has been considered since the nineteenth century as "normal" (Davis 1995). Making up new disease categories, or what medical science calls discovering new diseases, is a market-driven growth industry. "Health" is the absence of traits identified as disease, and "normal" is the statistical calculation of health. Normal has exceeded its original task of describing, ascending to become an aspirational moral ideal that Hacking (1990) calls "one of the most powerful ideological tools of the twentieth century" (169).

The practices and technologies of medical science aimed to make human biology, our very enfleshment, tractable by continually shaping and reshaping our future selves and communities to be ever healthier, always replacing the present with a better future. An abstract and ephemeral state of being, health is the desired achievement of modern medical science and those of us it shapes (Metzl and Kirkland 2010). Indeed, the legal theorist Dorothy Roberts (2010) notes that the achievement of health is "the unassailable aim of human biotechnologies" that "takes precedence over political and social interests" (61).

My friend now bears an existential burden about how this ideology of health would have operated to select preemptively against his now beloved son. One current example of how reproductive policy and practice enforces health by eliminating the birth of children with disabilities is Denmark's policy of testing all pregnant women for fetal Down syndrome. The number of babies born with Down syndrome has decreased significantly since testing began in the mid-1960s (Zhang 2020). In the name of promoting health, medical science has evolved an increasingly elaborate reproductive testing apparatus that identifies a growing number of human variations that women are asked to select for or against during pregnancy. This testing of what medical science understands as the *health* of the embryo or fetus can put the best interests of a pregnant woman in conflict with those of her future child. Such a complicated interaction of liberty interests can increase distress for pregnant women (Bernhardt et al. 2013; García, Timmermans, and van Leeuwen 2009; Samerski 2009; Stoll 2017; Werner-Lin, Mccoyd, and Bernhardt 2019). The 1973 Supreme Court decision extending reproductive liberty for women authorized this growth of reproductive testing and selection. In 1973, women's reproductive liberty consisted primarily of choosing whether to have children, but today, it consists of choosing the kind of children they will or will not have (Allyse and Michie 2022; Hvistendahl 2011; Watson 2018). The Supreme Court's overturning of its 1973 *Roe* decision with the 2022 *Dobbs* decision has intensified the complication of using prenatal testing for disability risk and selection to terminate pregnancies, a medical practice that Roe implicitly authorized in 1973 (Ziegler 2023).

The primary targets of reproductive selection in this recent testing-intensive reproductive environment are future persons predicted to have diseases, disabilities, or other traits that are considered disadvantages or burdens. This increased monitoring of embryos and fetuses yields abstract information such as risk profiles and prenatal diagnoses about a future child a woman has no access to knowing as a distinctive actual human being (Garland-Thomson 2022b). Her dilemma of needing to select for or against this hypothetical person creates a conflict between a woman's reproductive liberty and the right to life of future persons with disabilities. The phrase *right to life* has become a politically polarized concept in the United States in the legal and social conflict between an increasingly calcified pro- and anti-abortion movement. In an international context, however, a right to life for people with disabilities is a fundamental claim of the United Nations Convention on the Rights of Persons with Disabilities (UNCRPD). Adopted in 2006 and signed by 172 countries and the European Union, this international treaty protects people with disabilities and ensures a more extensive set of rights than the Americans with Disabilities Act by mandating support for people living with disabilities and affirming disability culture (United Nations General Assembly 2006).

Moreover, as science designs more ways to eradicate traits identified as diseases or disabilities from the genome, this conflict becomes more complex. The eugenic logic that enforces health as an unassailable aim takes precedence over ethical interests. For example, the 2020 report on Heritable Human Genome Editing, published by the National Academies of Sciences, Engineering, and Medicine (NASEM, 2020), calls for public debate about broader social and ethical questions and urges the use of genetic engineering only for serious diseases. Despite this, there is little opportunity for deliberation about what counts as a serious disease that goes beyond academic publications and gatherings of medical scientific experts (Baylis 2019; Sufian and Garland-Thomson 2021).

Scientific medicine is a taxonomy of human variation that counts most all human biological and psychological anomalies as a form of disease (Davis 2021; Rost 2021). In contrast, anomaly might be interpreted differently, through a lens of uncertainty or even unexpected benefit. Classifying a variation as a disease removes it from the context of the whole human being embedded in life, relationships, and community across time and space (Boardman and Clark 2022). Our collective understanding of the objective facts of disease symptoms and prognosis do not reliably map onto parental narratives of the entangled experiences of sorrow and joy, suffering and pleasure, and guilt and embrace that make up the lived experience of childhood suffering for a family during that experience and the continuing lifetimes such experiences shape. Conditions we might imagine as inflicting pain, shortening life, or reducing life quality are sometimes understood as tolerable or even beneficial, whereas conditions such as cystic fibrosis, which people can live with now for decades, are viewed by some families as a disease that produces unlivable lives (99 balloons 2023; Gann 2017).

Such reductive narratives that turn human traits into disease categories erase the complexity of lived lives of people bearing these labels and those who care about and for them. Defining health as the absence of human variations that count as disease constricts our imaginations about quality of life and human flourishing (Garland-Thomson 2019). Yet family narratives could reshape some policymaking and practice decisions as well as health-care training that might work toward more equitable or patient-centered understandings of medical treatment.

A New Eugenics

What does this meditation on eugenic science and its medical practice have to do with CRISPR, the newest and most promising tool in the suite of medical technology with which our fast-paced system of research, development, and commerce has presented us? Much of the public and professional conversation about CRISPR centers on explaining how it works, debating its safety, assessing its potential benefits, considering its targets, or warning against its unintended consequences. My concern is not with the efficacy or ingenuity of the technology, but rather with epistemological questions about what the existence of CRISPR technology suggests about the limits of being human—and what it means for my friend who someday might alter an embryo to align with what is considered a healthy child.

I have invoked the history of eugenics in modernity to support the position in the public and academic debates that much current reproductive technology, including gene editing, carries out a *new eugenics* in the name of health and reproductive liberty. The other side of the debate supports the free development and use of these reproductive technologies, often amplified by commercial interests. An ethics grounded in liberty interests strongly supports the growth of this laissez-faire medicine in today's moment when public sector or common good enterprises and private commercial interests are increasingly entangled. The commercial logic of free choice enters the obstetrical medical environment not only in support of reproductive liberty, but also in the name of a parental and medical obligation to fulfill the best interests of future children. For instance, a fetus, diagnosed by reproductive technology with spina bifida, can potentially receive in utero surgical treatment or be aborted, depending on the mother's exercise of her medical autonomy, within the limits of state law and local medical protocol. The burden of such a choice falls heavily on the mother trying to weigh the harms and benefits regarding the parental obligation to give one's child a good life. Many of these stories enter public conversation as books and articles about the complex network of suffering and joy as well as trouble and reward when a child with an unexpected medical condition or disability enters a family (Gann 2017; Rapp 2014; Solomon 2012). The opportunity to operate on the fetus is a choice a mother can make, but her choice is influenced by the opposing societal views of the fetus's future health versus the mother's reproductive freedom.

The ethical issues today's new eugenics bring forward concern the dynamics among correction, repair, improvement, and elimination as approaches to the development and use of medical technologies such as CRISPR. If the broadest ethical goal of any medical technology is to improve human lives, we must untangle some of the aspirations of eugenics from the enterprise of genetic technology and other medical interventions aimed at bringing all humans to a standard, "normal" form and function (Hacking 1990). Characteristics that depart from that standard in ways we understand as disadvantageous are human variations we think of as disease. Characteristics we understand as advantages that depart from that standard are often sought after as enhancements. Eugenics seeks to improve by eliminating the characteristics considered at a particular time and place to be disadvantages and to maximize those considered normal. Enhancement premises intensify the benefits of normal to create forms of super advantage. Genetic manipulation provides a seductive opportunity to improve society and individuals by bringing the abnormal toward normal and lifting the advantage of normal toward an intensified advantage of an imagined supernormal. Such a mechanical understanding of humans as compilations of individual characteristics that can be added or subtracted by way of medical intervention reduces us to the sum of our genetic profiles. Because the body–mind characteristics we think of as disease or disadvantageous traits are always parts of a whole living human being, snipping them away or fastening on supposedly better traits— to use the metaphors of editing and cutting employed to understand and explain CRISPR   promotes a crude understanding of human lived embodiment. The application of eugenic thinking in the first decades of the twentieth century ended because it failed to recognize that human beings could not simply be improved by wiping away specific characteristics deemed disadvantages from whole human beings embedded in lives and worlds.

A collective caution against the enthusiasm for this reductive understanding of improving human lives comes from historians such as Daniel Kevles (1985), bioethicists such as Nathaniel Comfort (2012, 2015), Nicholas Agar

(2010), Inmaculada de Melo-Martín (2017), and Françoise Baylis (2019), political theorists such as Michael Sandel (2007), and philosophers such as Jürgen Habermas (2003), who all argue against the liberal eugenics that genetic editing seeks to achieve. These thinkers hold that genetic manipulation for the enhancement or improvement of future persons or communities creates morally unacceptable consequences, ranging from producing medical harm to abrogating consent, intensifying genetic discrimination, increasing social inequality, promoting conditional parental acceptance, turning people into products, fostering a commercial medical industrial complex, and encouraging rogue scientific and medical practice. Many who oppose genetic editing understand it as scientific paternalism and a resource grab that saps funding from other initiatives that support the public good. Habermas speaks strongly for them all with the conclusion that genetic editing is "liberal eugenics regulated by supply and demand" (14).

Commercialized medical technology development in the interest of this liberal eugenics produces a culture of what de Melo-Martín (2017) calls *reprogenetics* that *standardizes human variation* in the interest of individual, market-driven liberty at the expense of social justice and the robust diversity and inclusion upon which modern egalitarian social orders depend. Such technology development and use go beyond genetic editing to a range of reproductive testing and selection practices that carry out what I call a *velvet eugenics* (Zhang 2020). Velvet eugenics takes its reference from the Velvet Revolution, beginning in 1989, that overturned many of the communist republics in Central and Eastern Europe without overt violence. Velvet as a metaphor suggests making a smooth change, using only the finest, commercially available product for the well-resourced consumer. This modern laissez-faire striving for what is understood by an individual at a specific time and place as the best drives much of the market for healthy conceptions, pregnancies, and curated offspring that for-profit genetic testing companies cultivate (Wasserman 2021).

By recognizing the eugenic work of medical science in the modern era, these historians, bioethicists, and philosophers offer a collective caution that recognizes the limits of the human capacity to control the future through actions in the present, no matter how well intended, carefully conceived, morally considered, or rigorously monitored (Barker and Wilson 2019; de Melo-Martín 1998; Garland-Thomson 2011; Knowles 2007; Sparrow 2011).

In opposition to these existential realists are techno-optimists, who cling to the conviction that the technologies medical science develops and uses can control outcomes beneficial to both future individuals and the human community. Sanguine futuristic aspirations, such as eliminating all human disease, enthusiastically supported by the psychologist Steven Pinker (2015), or creating a future population composed of what the philosophers Julian Savulescu and Guy Kahane (2009) call "the best," ignore or even dismiss both rogue uses of these eugenic technologies and unintended consequences. Such faith in what the twentieth century named as progress flies in the face of what the twenty-first century knows about the collateral damage ensuing from innovations ranging from nuclear energy to gasoline-powered engines to the ubiquity of plastic, sugary drinks, and opioid pain medication—all aimed at making a better future for everybody. Just as we collectively failed in the past to anticipate the future harms of what we took to be progressive benefits, many advocates of genetic manipulation technologies today refuse to consider the complexities of how and who these technologies may harm.

The Physics of Experience and the Limits of Human Knowledge

I return now to my friend and to the epistemological burden he carries. That burden—the perverse knowledge of the wisely deliberated, well-informed decision he would have made about his son's existence—he now understands as the worst choice he might ever have made. In facing that decision, my friend would have had an excess of one kind of knowledge—genetic, medical, and biological—and a deficiency of a different kind—social, cultural, and relational. Genetics, at its best, can predict only a broad range of outcomes. It may be presented to us as authoritative and reliably predictive knowledge, but it can be quite limited. In the focused context of the medical environment, however, genetics does have the advantage of being quantifiable and authoritative, whereas social and cultural knowledge is diffuse and ephemeral. The main limitation of genetic knowledge is that it tells us much less than it appears to predict. We may know that someone has what medical science determines to be the gene for what it has defined as a disease and is, therefore, likely at risk for or even certain to be shaped by it in some way. Yet such a preemptive diagnosis does not give us meaningful information about that person's capabilities, relationships, or actual lived life. Such genetic knowledge, like much aggregate knowledge drawn from statistical measurement, predicts little about how a particular lived life will unfold.

One way we might understand such a paradoxical relationship between our then self and now self—between what my friend knew then, what he knows now, and what he may know in the future—is to consider a phenomenological approach to the task of intentional family making and community making that new technologies such as CRISPR invite us to undertake. In other words, to think about the proper uses of CRISPR, we must recognize how the materiality of human biology and the physics of our placement in time and space structures human understanding and action. As both the philosophical tradition of phenomenology and modern science tell us, our human bodies, situated in moments and places, both shape and limit how we perceive knowledge and the action we take based on that knowledge. The historian Andrew Solomon (2019) affirms this received wisdom in his *New York Times* essay on "The Dignity of Disabled Lives." The lives of people with disabilities, Solomon recognizes, "are charged with inherent dignity." This dignity manifests in how we carry out a life shaped by the embodied particularities the world counts as disabilities. Offering his own life of flourishing as an example, Solomon claims, "We are mostly an accumulation. . . . If I imagine myself without dyslexia, without A.D.D., without depression, without gayness, without nearsightedness, without orthostatic hypotension, without Jewishness, without white privilege, without prosopagnosia, then there's very little of me left" (para 26). This physics of human experience perhaps makes medical technologies such as CRISPR adversaries rather than friends of human flourishing—not because they will not work properly, but because they exist beyond the limits of human knowledge and imagination. The development and deployment of the actual material phenomenon that scientists describe as CRISPR-*Cas*9 is outside of direct human experience. Moreover, its development and implementation exist outside of the context of lived human lives embedded in specific times, places, relationships, and communities. Like black holes, the scale of the material substance we call *genes* requires humans to deduce its existence and nature through indirect observation and scientific experiments involving the differential results of intervention. As with the development of artificial intelligence, nuclear energy, weapons of mass destruction, and technology in general, we need to approach our investment and use of them with humility and respect for the vastness of human particularity and experience.

Human understanding also takes shape through the ways that language structures thought and prompts action. We use metaphors that compare one thing to another to know our world (Lakoff and Johnson 1980). Metaphors such as gene editing give us a picture of machines such as typewriters or computers doing routine mechanical work that our everyday experience assures us is stable and safe. Microsurgery inserting biological bits into other biological bits is on a scale and location unavailable to actual human perception and experience. To compare our everyday use of machines like computers to a surgical process carried out deep inside the human body can mislead us about the nature of the actual process we are calling "gene editing." Using mechanical metaphors to understand medical processes does the rhetorical work of strengthening the hubris that techno-optimism fundamentally leans toward. Such risk of overconfidence can shape the moral and ethical decisions we make about using CRISPR (O'Keefe et al. 2015). We need to recognize that the physics of our human biology constitutes a constraint on our understanding. As Kircher et al. (2014) write: "Our capacity to sequence human genomes has exceeded our ability to interpret genetic variation" (310). Such an acknowledgment of human phenomenological and epistemological limitations can lead us to an ethically necessary humility about the extent of human control over the consequences of the enterprise we think of as human gene editing. To honor human limitation would also honor the ethical commitment to humility ensconced in the creed to "first, do no harm." This first principle of medicine and bioethics elevates the caution of nonmaleficence over the aspiration to beneficence (Beauchamp and Childress 2012). Before we consider the benefits of our eager strides to develop and implement technologies that shape the outcomes of human reproduction and the future persons included in our human communities, we may be wiser to carefully consider the harms.

The Wisdom of Humility

If this distinctly modern aspiration to control the future through rational, intentional actions in the present is as fragile as the existential thinkers feared, and as the liberal eugenicists and techno-optimists deny, how should we understand our relationships with and obligations to future generations? In other words, what might be an alternative to the eugenic future I described here that haunts my friend? What can we put up against medical science's compelling progress narrative about developing and using tools to shape better future populations without what we consider diseases or disabilities?

To contemplate an alternative to the logic of a eugenic future, I draw from the wisdom of two famous scientists, both of whom contributed significant scientific knowledge and offered abundant meditations on what it means to be human. In their prolific public writing, Stephen Jay Gould, a paleontologist at Harvard, and Oliver Sacks, a neurologist from Oxford, studied us and told us over and over, with more wonder than dismay, what they saw us to be (Gould 1981, 1985; Sacks 1985, 1995, 2015). For decades, both men wrote with increasing urgency in our most influential venues for public conversation. As each of them was dying from cancer, they imparted their final words, conclusions drawn from a lifetime of scientifically observing their fellow humans. Gould (1985) reflected on his cancer diagnosis by explaining the statistical calculations doctors use to predict life expectancy for cancer patients, clarifying how individual variations cannot be accounted for in the narrative of futurity that medical science calls prognosis. In pondering the stability between predicted and actual outcomes, Gould's scientific expertise in evolution led him to conclude that "variation itself is nature's only irreducible essence" (40). Nature's essence, Gould reckons, is contingency, a predictable unpredictability. To standardize human variation through medical interventions thought to improve, normalize, or make healthy future persons might be not only imprudent, but also a kind of hubris. Such shaping of future generations is the Promethean overreach that the political theorist Michael Sandel (2007) finds in the aspiration to use medical technologies that make us into what present persons prefer or think is best.

Oliver Sacks's late wisdom, like Gould's scientific insight, had its beginnings in his clinical observations of patients with rare neurological disorders. His case studies wavered between detailing his observations of patients with medical diagnoses such as encephalitis or Tourette's syndrome to abandoning pathological classifications in favor of narrative accounts of individual human perception and action such as *The Man Who Mistook His Wife for a Hat*. Moving between medical reports and literary chronicles in his descriptions of his patients, Sacks captured the distinctiveness of these unusual ways of being in the world. He shifted from reporting on clinical case studies to collecting ethnographies of the exotic people to whom he was persistently drawn. Toward the end of his life, especially after his cancer diagnosis, he became his own case study, finally a patient akin to the ones he had been observing from the beginning. His final wisdom appeared in a series of introspective personal essays in the *New York Times*, collected posthumously under the appropriate title *Gratitude*. Anticipating his own imminent death, Sacks concluded, "When people die, they cannot be replaced. They leave holes that cannot be filled, for it is the fate—the genetic and neural fate—of every human being to be a unique individual, to find his own path, to live his own life, to die his own death" (Sacks 2015).

Both Gould and Sacks determine similarly that the biological variations inflecting our fundamental sameness shape our social and evolutionary destiny. This truth is both philosophical and scientific: genotype diversity is not only the essence of life, but phenotype diversity, our distinctive human individuality, gives life meaning. The wisdom of these philosopher-scientists counsels us that the intricate intertwining of sameness and difference that links human generations is an ever-emerging biological and social system we should not endeavor to control, but rather over which we should maintain a respectful stewardship. This is the scientific and humanistic truth, of course, that underpins my friend's conviction that parenting the *unique individual* embedded in the network of human relationships that his son has become has given meaning to his life.

A Call for Humane Technologies

How might such a claim from these contemporary wise men square with the troubled human history of gathering ourselves into tribes of imagined sameness and turning upon those we imagine as different from us? Why have we organized ourselves in communities ever suspicious of human variation, ever ready to recognize one another as foreigners and attack each other on that account (Glover 2001)? Such questions underlie how we have organized ourselves into communities across time and place. The paradox of our simultaneous requirement for and intolerance of human variation is, perhaps, the fundamental challenge for our human communities in living and navigating the future together (Arendt 1958).

Considering what Solomon, Gould, and Sacks have observed about the value of distinctive embodied human lives lived out rather than preordained by the preferences of others, we can conclude that eugenic science, whether supported by totalitarian or liberal regimes, runs counter to the best interests of human communities and individuals. The wisdom from these recorders of the human condition suggests that we distinguish between eugenic and humane technologies (Wilson 2018). The challenge is to recognize that any technology only becomes eugenic or humane depending on how we employ it (Bergstresser 2022). To consider how we use human knowledge and capability, we might turn to insights from the archive of cautionary tales our inherited traditions offer—from Adam and Eve in their Garden, Prometheus chained to his rock, Pandora before her box, to Oedipus fleeing from his fate. These traditional stories gather lived human experience to render how our intentions and intended outcomes entangle in unexpected ways as we carry out the projects of our lives. Attention to this received wisdom might serve us well in considering the development and use of technologies today.

Now and throughout the human past, eugenic technologies aim to control contingency by medically producing future persons and communities according to the values of ascendant groups at certain times and places in human history. In contrast, humane technologies honor and accommodate the widest range of human variation. That means designing and building a world shaped to fit human variation and individual distinctiveness, rather than attempting to shape bodies to fit into the extant world (see Hamraie 2017, Hendren 2020, Williamson and Guffey 2020). It also means developing an attitude of humility toward human biodiversity when imagining another's life and how any of us live (Garland-Thomson 2022a). This requires resisting a notion of technological progress that continually categorizes human diversity in terms of new disease to be identified and eliminated. Such an enterprise regularizes human minds and bodies according to concepts of advantage or disadvantage, a narrow version of health, a commercially incentivized growth industry of pathological diagnosis, or a notion of individual liberty at the expense of the common good (Garland-Thomson 2012). Humane technologies, in contrast, support the existence of present and future persons in our human distinctiveness.

Humane technologies, then, create an environment in which people grow into themselves (Habermas 2003). For humans to thrive, we need to be ensconced in an environment that sustains the form, function, and needs of our bodies. A sustaining environment in which we can flourish provides sturdy and reliable human connections, access to adequate economic and material resources, strong community networks, protection from violence and harm, and a humane form of social organization. We should devote technology to the humane goal of providing every human being with a genuinely open future to grow into a distinct, singular individual, rather than making them as we think they should be. Providing this open future does not preclude appropriate interventions—sometimes medical, sometimes social—that ameliorate pain and suffering, promote human functioning and flourishing, or provide minimally invasive technologies that bridge flesh and world (Stramondo 2020). Rather than researching and developing commercial medical technology to reduce human variation, we should invest more resources in sustaining and supporting human variation as it exists. Resource allocation should develop and maintain environments and technologies through which human distinctiveness can flourish. Within a humane technologies framework, liberty can be understood as the freedom to grow from our distinctive individuality, not according to conceptions of health, normalcy, advantage, preferences, monetary cost, or future concepts of life quality imposed through parental will or medical authority and justified as the best interests of the future person. Humane technologies accommodate rather than eliminate human diversity.

We must understand that the technologies humans develop can benefit or harm our human communities, depending upon how we use them. Any technology, including genetic-editing technologies, can be used to both benefit and harm human individuals and their communities. How we implement, both in the present and the future, any technology is what makes it humane or eugenic. The challenge for us is to distinguish between how and when we use any tools we make for ethical humane use or to harm one another and our shared world. In developing, recognizing, and implementing humane technologies, we should remember that we do not know with certainty what counts as benefit or harm, advantage or disadvantage in the long arc of human history. "The purpose of evolution," geneticist and Nobel Laureate Mario Capecchi told me, "is to anticipate the unexpected" (personal communication, July 31, 2017). Human variation moves evolution forward, yielding new forms that are fresh solutions to changing environments both natural and human designed. Characteristics we think of today as health, intelligence, or strength may not serve future generations in the environments or life tasks they must undertake, just as such imagined traits are no guarantee of flourishing for any of us now. Relying on CRISPR to shape the future, according to the perspectives and intentions of the present by eliminating what we think of as *disabilities*, presumes we know our future and how best to construct human beings to face challenges we cannot fully imagine.

Even a brief survey of human history—and certainly recent debates over medical technologies involving gene editing—suggests that too many of us assume what living with the human variations we think of as disabilities must be like. These assumptions often come from a faulty comparison between how some of us live with our body-minds considered normal and how others of us live with body-minds considered disabled. Overdetermined diagnostic categories described in terms of syndrome, inadequacy, retardation, deformity, fatal, or incurable further divide humanity into opposing and often calcified categories such as normal and abnormal, healthy and disabled, or thriving and doomed. Rather than seeking to change the future lives of our fellow humans we might imagine as different from us, we may do better to find a wider range of human stories to tell and to shape our collective decisions about how we make and use technologies such as CRISPR to eliminate them.

Sacks and Gould, as well as my friend with the beloved son, remind us that the short span of human life and imagination limits our capacities to anticipate the consequences of present actions to control future outcomes. To do its work of curing, medical science conceptualizes future persons as composites of traits identified as normal or abnormal, rather than imagining individuals like my friend's son as the whole person his father experiences him as being. CRISPR has the potential to reach far across the long arc of human history in ways that we today cannot control or imagine. CRISPR's eugenic promises blunt our capacity to appreciate the unexpected. From loving his son and from the relationships that love has secured, my friend has learned that the human variations we think of as disabilities or disease are inseparable from the wholeness of our being. "All creatures who persist are whole," one storyteller reminds us about disabled lives lived well (Lambeth 2016). What my friend knows now about his son is that this persistence is one source of great dignity.

REFERENCES

Agar, Nicholas. 2010. *Humanity's End: Why We Should Reject Radical Enhancement*. Cambridge: MIT Press.

Allyse, Megan A., and Marsha Michie, eds. 2022. *Born Well: Prenatal Genetics and the Future of Having Children*. New York: Springer.

Arendt, Hannah. 1958. *The Human Condition*. Chicago: University of Chicago Press.

Barker, Matthew J., and Robert A. Wilson. 2019. "Well-Being, Disability, and Choosing Children." *Mind* 128 (510): 305–28.

Baylis, Françoise. 2019. *Altered Inheritance: CRISPR and the Ethics of Human Genome Editing*. Cambridge: Harvard University Press.

Beauchamp, Tom L., and James F. Childress. 2012. *Principles of Biomedical Ethics*. 7th ed. New York: Oxford University Press.

Bergstresser, Sara M. 2022. "Eugenic Technologies Are Developed in Eugenic Eras: Why We Must Include Historical Circumstances in Socio-Political Perspectives for Neuroethics." *AJOB Neurosci* 13 (1): 28–30. https://doi.org/10.1080/21507740.2021.2001091.

Bernhardt, Barbara A., Danielle Soucier, Karen Hanson, Melissa S. Savage, Laird Jackson, and Ronald J. Wapner. 2013. "Women's Experiences Receiving Abnormal Prenatal Chromosomal Microarray Testing Results." *Genet Med* 15 (2): 139–45.

Boardman, Felicity K., and Corinna C. Clark. 2022. "What Is a 'Serious' Genetic Condition? The Perceptions of People Living with Genetic Conditions." *Eur J Hum Genet*, 30 (2): 160–69. https://doi.org/10.1038/s41431-021-00962-2.

Cohen, Adam. 2016. *Imbeciles: The Supreme Court, American Eugenics, and the Sterilization of Carrie Buck*. New York: Penguin.

Comfort, Nathaniel. 2012. *The Science of Human Perfection: How Genes Became the Heart of American Medicine*. New Haven: Yale University Press.

Comfort, Nathaniel. 2015. "Can We Cure Genetic Diseases without Slipping into Eugenics?" *Nation*, July 16, 2015. https://www.thenation.com/article/archive/can-we-cure-genetic-diseases-without-slipping-into-eugenics/.

Davis, Joseph E. 2021. "All Pathology, All the Time." *New Atlantis* 66: 55–65.

Davis, Lennard J. 1995. *Enforcing Normalcy: Disability, Deafness, and the Body*. New York: Verso.

de Melo-Martín, Inmaculada. 1998. *Making Babies: Biomedical Technologies, Reproductive Ethics, and Public Policy*. Boston: Kluwer Academic.

de Melo-Martín, Inmaculada. 2017. *Rethinking Reprogenetics: Enhancing Ethical Analyses of Reprogenetic Technologies*. New York: Oxford University Press.

Duster, Troy. 1990. *Backdoor to Eugenics*. New York: Routledge.

Fries, Kenny. 2017. "The Nazis' First Victims Were the Disabled." *New York Times*, September 13, 2017. https://www.nytimes.com/2017/09/13/opinion/nazis-holocaust-disabled.html.

Gann, Jennifer. 2017. "Every Parent Wants to Protect Their Child. I Never Got the Chance." *Cut*, November 26, 2017. https://www.thecut.com/2017/11/raising-child-with-cystic-fibrosis.html.

García, Elsa, Danielle R. M. Timmermans, and Evert van Leeuwen. 2009. "Reconsidering Prenatal Screening: An Empirical-Ethical Approach to Understand Moral Dilemmas as a Question of Personal Preferences." *J Med Ethics* 35 (7): 410–14.

Garland-Thomson, Rosemarie. 2007. "Transferred to an Unknown Location." *Disabil Stud Q* 27 (4).

Garland-Thomson, Rosemarie. 2011. "Human Biodiversity Conservation: A Consensual Ethical Principle." *Am J Bioeth* 15 (6): 13–15.

Garland-Thomson, Rosemarie. 2012. "The Case for Conserving Disability." *J Bioeth Inq* 9 (3): 339–55. https://doi.org/10.1007/s11673-012-9380-0.

Garland-Thomson, Rosemarie. 2017. "My Orphan Disease Has Given Me a New Family." *New York Times*, October 26, 2017. https://www.nytimes.com/2017/10/26/opinion/my-orphan-disease-has-given-me-a-new-family.html.

Garland-Thomson, Rosemarie. 2019. "Welcoming the Unexpected." In *Human Flourishing in an Age of Gene Editing*, edited by Erik Parens and Josephine Johnston, 15–28. New York: Oxford University Press. https://doi.org/10.1093/oso/9780190940362.003.0002.

Garland-Thomson, Rosemarie. 2022a. "Evaluating the Lives of Others." *Am J Bioeth* 22 (9): 30–33. https://doi.org/10.1080/15265161.2022.2105607.

Garland-Thomson, Rosemarie. 2022b. "The Hypothetical Healthy Newborn." In *Born Well: Prenatal Genomics and the Future of Having Children*, edited by Megan A. Allyse and Marsha Michie. New York: Springer.

Glover, Jonathan. 2001. *Humanity: A Moral History of the Twentieth Century*. New Haven: Yale University Press.

Gould, Stephen Jay. 1981. *The Mismeasure of Man*. New York: Norton.

Gould, Stephen Jay. 1985. "The Median Isn't the Message." *Discover* 6 (6): 40–42.

Habermas, Jürgen. 2003. *The Future of Human Nature*. Cambridge: Polity.

Hacking, Ian. 1990. *The Taming of Chance*. Cambridge: Cambridge University Press.

Hacking, Ian. 2006. "Making Up People." *London Rev Books* 28 (16). https://www.lrb.co.uk/v28/n16/ian-hacking/making-up-people.

Hamraie, Aimi. 2017. *Building Access: Universal Design and the Politics of Disability*. Minneapolis: University of Minnesota Press.

Hendren, Sara. 2020. *What Can a Body Do?* Riverhead Books: New York.

Hvistendahl, Mara. 2011. *Unnatural Selection: Choosing Boys Over Girls, and the Consequences of a World Full of Men*. New York: PublicAffairs.

Kevles, Daniel J. 1985. *In the Name of Eugenics: Genetics and the Uses of Human Heredity*. New York: Knopf.

Kircher, Martin, Daniela M. Witten, Preti Jain, Brian J. O'Roak, Gregory M. Cooper, and Jay Shendure. 2014. "A General Framework for Estimating the Relative Pathogenicity of Human Genetic Variants." *Nat Genet* 46 (3): 310–15.

Knowles, Lori P. 2007. *Reprogenetics: Law, Policy, and Ethical Issues*. Baltimore: Johns Hopkins University Press.

Lakoff, George, and Mark Johnson. 1980. *Metaphors We Live by*. Chicago: University of Chicago Press.

Lambeth, Laurie Clements. 2016. "The Three-Legged Dog Who Carried Me." *New York Times*, December 7, 2016. https://www.nytimes.com/2016/12/07/opinion/the-three-legged-dog-who-carried-me.html.

Lombardo, Paul A. 2008. *Three Generations, No Imbeciles: Eugenics, the Supreme Court, and Buck v. Bell*. Baltimore: Johns Hopkins University Press.

Metzl, Jonathan M., and Anna Kirkland, eds. 2010. *Against Health*. New York: NYU Press.

National Academies of Sciences, Engineering, and Medicine (NASEM). 2020. *Heritable Human Genome Editing*. Washington, DC: National Academies Press.

99 Balloons. 2023. "Changing the Story of Disability." https://99balloons.org/.

O'Keefe, Meaghan, Sarah Perrault, Jodi Halpern, Lisa Ikemoto, and Mark Yarborough. 2015. "'Editing' Genes: A Case Study about How Language Matters in Bioethics." *Am J Bioeth* 15 (12): 3–10.

Paul, Diane B. 1995. *Controlling Human Heredity: 1865 to the Present*. New Jersey: Humanities Press.

Pinker, Steven. 2015. "The Moral Imperative for Bioethics." *Boston Globe*, August 1, 2015. https://www.bostonglobe.com/opinion/2015/07/31/the-moral-imperative-for-bioethics/JmEkoyzlTAu90QV76JrK9N/story.html.

Proctor, Robert N. 1988. *Racial Hygiene: Medicine under the Nazis*. Cambridge: Harvard University Press.

Rapp, Emily. 2014. "'Dragon Mother' Emily Rapp: A New Baby Doesn't, and Shouldn't, Replace What's Lost." *New York Times*, March 13, 2014. https://archive.nytimes.com/parenting.blogs.nytimes.com/2014/03/13/dragon-mother-emily-rapp-a-new-baby-doesnt-and-shouldnt-replace-whats-lost/.

Roberts, Dorothy E. 2010. "The Social Immorality of Health in the Gene Age: Race, Disability, and Inequality." In *Against Health*, edited by Jonathan M. Metzl and Anna Kirkland, 61–71. New York: NYU Press.

Rost, Michael. 2021. "To Normalize is to Impose a Requirement on an Existence." Why Health Professionals Should Think Twice before Using the Term "Normal" with Patients. *J Bioeth Inq*, 18 (3), 389–94. https://doi.org/10.1007/s11673-021-10122-2.

Sacks, Oliver. 1985. *The Man Who Mistook His Wife for a Hat*. New York: Summit Books.

Sacks, Oliver. 1995. *An Anthropologist on Mars*. New York: Knopf.

Sacks, Oliver. 2015. "My Own Life." *New York Times*, February 19, 2015. https://www
.nytimes.com/2015/02/19/opinion/oliver-sacks-on-learning-he-has-terminal-cancer
.html.

Samerski, Silja. 2009. "Genetic Counseling and the Fiction of Choice: Taught Self-
Determination as a New Technique of Social Engineering." *Signs* 34 (9): 735–61.

Sandel, Michael J. 2007. *The Case Against Perfection: Ethics in the Age of Genetic Engineering*.
Cambridge: Harvard University Press.

Savulescu, Julian, and Guy Kahane. 2009. "The Moral Obligation to Create Children with
the Best Chance of the Best Life." *Bioethics* 23 (5): 274–90.

Solomon, Andrew. 2012. *Far from the Tree*. New York: Scribner.

Solomon, Andrew. 2019. "The Dignity of Disabled Lives." *New York Times*, September 2,
2019. https://www.nytimes.com/2019/09/02/opinion/disabled-human-rights.html.

Sparrow, Robert. 2011. "A Not-so-New Eugenics: Harris and Savulescu on Human
Enhancement." *Hastings Cent Rep* 41 (1): 32–42.

Stoll, Katie. 2017. "The Routinization of Prenatal Testing and the Erosion of Patient Autonomy." *DNA Exchange*, November 16, 2017. https://thednaexchange.com/2017/11/16/the
-routinization-of-prenatal-testing-and-the-erosion-of-patient-autonomy/.

Stramondo, Joseph A. 2020. "Disability and the Damaging Master Narrative of an Open
Future." *Hastings Cent Rep* 50, S30–36. https://doi.org/10.1002/hast.1153.

Sufian, Sandy, and Rosemarie Garland-Thomson. 2021. "The Dark Side of CRISPR."
Scientific American, February 16, 2021. https://www.scientificamerican.com/article
/the-dark-side-of-crispr/.

United Nations General Assembly. 2006. *Convention on the Rights of Persons with Disabilities*. New York: United Nations, A/RES/61/106. December 13, 2006. https://www
.un.org/development/desa/disabilities/convention-on-the-rights-of-persons-with
-disabilities.html.

Wasserman, David. 2021. "An Intelligent Parents Guide to Prenatal Testing: Having a Well-
Born Child Without Genomic Selection." In *Born Well: Prenatal Genetics and the Future
of Having Children*, edited by Megan A. Allyse and Marsha Michie, 125–36. New York:
Springer.

Watson, Katie. 2018. *Scarlet A: The Ethics, Law, and Politics of Ordinary Abortion*. New York:
Oxford.

Werner-Lin, Allison, Judith L. M. Mccoyd, and Barbara A. Bernhardt. 2019. "Actions and
Uncertainty: How Prenatally Diagnosed Variants of Uncertain Significance Become
Actionable." *Hastings Cent Rep* 49 (3): S61–71. https://doi.org/10.1002/hast.1018.

Williamson, Bess, and Elizabeth Guffey, eds. 2020. *Making Disability Modern: Design Histories*.
London: Bloomsbury Visual Arts.

Wilson, Robert A. 2018. "The Eugenic Mind Project." April 17, 2018. http://eugenicsarchive
.ca/discover/timeline/5ad67ea98eacc7685a00000a.

Ziegler, Mary. 2023. *Roe: The History of a National Obsession*. New Haven: Yale University
Press.

Zhang, Sarah. 2020. "The Last Children of Down Syndrome." *Atlantic*, December 2020.
https://www.theatlantic.com/magazine/archive/2020/12/the-last-children-of-down
-syndrome/616928/.

12

A Therapeutic Fallacy

Peter F. R. Mills

Among the objections to the implementation of heritable genome editing (HGE) is that it does not address an unmet medical need and, therefore, fails to meet an important criterion for introducing an unproven procedure with uncertain and potentially adverse consequences. The objection mostly lies in the potential harms of the clinical use of HGE, such as unpredicted negative consequences resulting from "off-target" mutations or from editing a gene with multiple unaccounted-for functions. Other adverse effects could be social, like the disturbance of valued interpersonal and intergenerational norms through the commodification of children's inherited characteristics and the normalization of interventions to secure preferred phenotypes. Nevertheless, much of what has been said and written about HGE focuses on the possibility of making it safe enough for use in a clinical setting to respond to perceived unmet medical needs. Therefore, we must turn our attention first to the question of whether HGE satisfies an unmet medical need. In the absence of a pressing medical need, the potential for deleterious outcomes would create a strong presumption against using HGE that may be difficult or impossible to overcome.

Are there Unmet Needs?

Part of the objection to HGE is that for all the possible outcomes it may be expected to deliver, reproductive technologies already exist that can secure them. These outcomes are usually assumed to be the birth of a child with at least two desired characteristics. The first desired characteristic is a direct genetic link to both parents. This link is usually achieved through the exclusive combination of the parental genomes that occurs in unassisted sexual reproduction. The second characteristic is the absence of at least one genetic feature any child resulting from unassisted reproduction may inherit. This might be the absence of a serious disease, like cystic fibrosis, that has been diagnosed in the prospective parents. Genetically transmitted disease traits of this kind are usually assumed to be good candidates for HGE.

In cases where genetic diseases may be passed from a parent or parents to their offspring, HGE is usually not the only intervention available. Well-established reproductive technologies that use sperm or eggs from screened, third-party donors can prevent the transmission of deleterious genetic traits from one or both parents. Because these children would lack the first pre-ferred characteristic of genetic relatedness, other approaches are also widely available in many (though not in all) parts of the world. These include gene-tic testing to select embryos for transfer—preimplantation genetic diagnosis (PGD)—or prenatal testing and termination of pregnancy (PNT/ToP). Highlighting these alternative reproductive strategies helps to differentiate between technical and subjective appraisals of what constitutes an acceptable alternative to PGD or PNT. I will explore technical questions first.

It is true that existing approaches that help people who carry genes for a variety of inherited diseases to have unaffected children are subject to their own distinctive limitations. For example, with PGD and PNT, the limited number of embryos available may mean that the chances of finding an em-bryo or pregnancy with the preferred characteristics may be small. They are further reduced if the aim is to find an embryo with *two* characteristics that assort independently. These might include the absence of an inherited con-dition such as beta thalassemia and the capability to act as an HLA-compatible tissue donor to treat a family member already affected by that condition. It is notable that in such a rare case, the parental link that makes this tissue compatibility more likely may strengthen the case for using HGE, because a child conceived using an unrelated egg or sperm would be much less likely to be a compatible tissue donor. Nevertheless, conceiving a child to have a com-patible tissue donor for another child or parent raises complex ethical ques-tions, including novel implications for familial relations. Added to these limited probabilities are the uncertainties and burdens of repeated cycles of in vitro fertilization (IVF) treatment, which are needed for PGD and HGE, and the broader impact of these treatments on individuals and families.

There are, however, some highly unusual cases in which none of the es-tablished reproductive alternatives is feasible because offspring would *always* inherit their parents' genetic condition. These unusual cases include couples where one person is homozygous (has two identical copies of a gene) for an autosomal dominant condition (such as homozygous familial hypercholester-olemia) or where both are homozygous for a recessive one (such as sickle cell disease). Though rarely seen in clinical practice, these cases are not un-known and, despite the health challenges these people face themselves, it is not unreasonable that they should wish to become parents. It is these "most unusual cases" in which PGD or PNT offer no route to having a child with the two preferred characteristics (because all the parental embryos will carry a particular, but rare, genetic disease) that are often imagined to offer the best argument for using HGE.

A Therapeutic Fallacy

These very uncommon situations, however, may still fail to meet the criterion of "unmet medical need." Certainly, if HGE were the "last best hope" of averting an imminent harm to a *living* individual, compassion demands that it should be considered seriously. Such an approach was relied upon in the case of the first person to be treated using genome-edited cells—a baby girl with relapsed acute lymphoblastic leukemia, successfully treated with modified CAR-T cells at the Great Ormond Street Hospital in London in 2015 (Qasim et al. 2015). That kind of situation, however, is plainly not the case with HGE. What is at stake, first and foremost, is assisting a given couple to have a child. While the chosen approach certainly involves some complex medical procedures, it would be a fallacy to assume that simply because HGE involves a medical procedure, it necessarily has a therapeutic purpose.

But is it correct to say that prospective parents at risk of passing on a ge-netic condition to their future offspring do not have a medical need? While some genetic conditions may be contingently associated with clinical infer-tility, assisted conception strategies characteristically circumvent, rather than treat, underlying infertility. It is widely acknowledged that the inabil-ity to conceive a child (or a child with or without certain features) can be a profoundly felt need that raises personal and sensitive questions and, for those reasons, is often exempted from interrogation. But it is also widely ac-cepted that reproductive rights cannot be so unqualified as to co-opt the services of clinical professionals as if they were mere instruments for secur-ing the interests of prospective parents and their future children. For the time being, then, let me note that "need" expresses an interest of surpassing importance for the individuals involved and one that is often widely endorsed by others. It might even be so profound as to affect the mental or physical health of people experiencing it in ways comparable to other forms of disease and disability. To treat it as a medical need on the part of prospective parents, however, would seem to cast the child they desire to have in a curiously instrumental role: a means rather than an end.

The prospective parents' need for a child is, then, relative to other important interests, most obviously the anticipated interests of their future child. Are we not also bound to consider what that child might need? Obviously, future children cannot speak for themselves, but let us make the reasonable assumption that the child will need, as a minimum, the conditions of a life worth living. No doubt most people would go further than this minimal condition: proposing to have a child who would enjoy a life barely worth living seems a poor parenting ambition (Parfit 1984). A "good enough" life, then, or conditions that are adequate in the circumstances, may be more a reasonable expectation (O'Neill 1979). This is where HGE comes in: it is assumed that HGE can ameliorate these life conditions, particularly in cases in which they would not otherwise meet the standard of a "good enough" life. It could, however, also fail to do so. More troublingly, it could have an adverse iatrogenic effect, making the child's life conditions worse. Which of the possible outcomes will play out, based on current knowledge, is still uncertain.

Does a prospective child at risk of being born with a genetic condition, then, not have a "medical need"? Many genetic variants give rise to phenotypic states that are diagnosed medically. Some of these are treated by medical interventions. But the question here is not about *treating* a condition diagnosed adventitiously but about *planning to bring about the birth* of a person with particular characteristics or one likely to be affected by such a condition. Even if we are prepared to say that the parents have a "need" for their child to exist, that child does not, before they are born, have a need to exist. In the context of the moral decision, the child is a *prolepsis*, an artifact of the imagination. We might conclude that if the child does not exist, then that child can, *a fortiori*, have no further needs. If, on the other hand, the child does exist, then they will "need" the sufficient conditions of a "good enough" life. Because there is some unavoidable uncertainty about whether the child will have these conditions for a good enough life because of HGE, it will always be necessary to weigh this possibility against the parents' "need" to have a genetically related child.

What makes all the difference here are the circumstances in which the course of action is determined. In the case of a disease diagnosed after birth, parents can respond only to the unanticipated discovery that their child—another living human being—has a deleterious condition. In the case of using HGE, prospective parents, and the clinicians who assist them, must assume responsibility for choosing the characteristics of a child they plan to bring into existence and for executing that plan, with all the risks and uncertainties attending it. These circumstances are novel. The situation in which prospective parents are furnished with information about the genetic implications of conceiving a child and have the technological possibility of altering the outcome by the deliberate intervention in the genome is unknown historically. If we peer into the future, however, we can imagine that even the choice to refuse this knowledge (for example, to decline preconception testing) or to reject the opportunity to intervene may impose a moral burden on prospective parents. They may even become subject to societal pressure to ensure that their child is free from genetic disorders.

The foregoing reflections are sufficient to show, I believe, why the model of experimental medicine—which informs much of the literature on pathways for using HGE—is inappropriate for introducing a radically innovative reproductive technology like HGE. Once used, HGE cannot be withdrawn or titrated. Most importantly, the clinical trial model places too much weight on the patients' consent and interests, and it does not integrate the difficult question of the future person's interests, independent from the prospective parents' interests in their child's existence. The prospective parents' consent (specifically that of the prospective mother) is primarily to a treatment that aims to give her a child. Even if the prospective child could somehow be said to have medical needs, those needs are conditional upon the fulfilment of their parents' (arguably) nonmedical needs: the child's need cannot make the HGE intervention necessary for the simple and sufficient reason that the child's existence is not necessary. On the contrary, the child's existence depends on the choice of the parents to use technology to bring that child into existence.

Medical Need and the Responsible Pathway

Inscribing HGE within a medical model has certain consequences. That the aim of the intervention is purported to be "medical" does not, strictly, strengthen the concept of "need," although it suggests a need that has authoritative and objective value (for example, in contrast to a subjective preference, desire, or whim). Its more obvious function, though, is to imply that the need deserves a distinctive response. Placing the need within a medical context may imply, for example, that it demands the attention of medical specialists, the allocation of medical resources, the delivery of interventions through medical institutions, and the application of medical governance arrangements, according to medical norms. The medical framing also suggests characteristic standards for valuing interventions (in relation to how they treat disease and promote health), appraising innovations (in terms of relative probabilities of benefit and harm), and respecting the autonomy of patients. These conditions provide important standards of good practice. The medical framing may also serve to arrogate the authority to determine when the conditions for HGE are met to a particular cadre of professionals. It brings within the competence of technical specialties the task of pronouncing when, in what circumstances, and under whose direction the risks are sufficiently low to proceed.

A polarization has emerged over the last few years between those who want to forge ahead with HGE, who see the primary obstacles to be overcome as technical, and those who see the most immediate obstacles as moral ones. The first group speaks in terms of a pathway along which technology can be trusted to develop through regulated steps, to the point where the question becomes not "why?" but "why not?" The second group speaks in terms of the need for a broad moral consensus or, at least, explicit consideration of what international differences of approach can be tolerated, before any such steps are taken (Baltimore et al. 2018; Lander et al. 2019).

After a few years of hedging, the proponents of moving forward with CRISPR have been goaded into setting out the prospective translational pathway for HGE. An example was offered in the 2020 "Consensus Study" Report of the International Commission on the Clinical Use of Human Germline Genome Editing convened by the US National Academies of Medicine and Sciences and the UK's Royal Society (NAMSRS 2020). To its credit, the International Commission report follows the earlier National Academies of Sciences, Engineering, and Medicine (NASEM 2017) report in recognizing the primacy of genetic relatedness as a condition for the uses of HGE it addresses. It sketches out six categories of cases (denoted A to F) in which HGE might be used. Apart from the broad "category E" cases, all of these involve heritable disease or disability. Only categories A and B, however, (where either none or too few unaffected embryos are available) are considered suitable for translational uses of HGE. This condition may also be met by Category C cases, but these are ruled out because the genetic conditions are potentially treatable or "less serious," like heterozygous familial hypercholesterolemia (which is considered manageable) and inherited deafness (which is considered less serious). The logic here is elliptical, though. It no longer revolves around only the parents' desires and preferences but has a second focal point, a quasi-objective medical judgment about conditions to be avoided in offspring that is about their seriousness rather than the likelihood of occurrence.

From the viewpoint of any future child, or of those who stand as stewards for the interests of future offspring, this is, as they say, curious. Consider: if the genome editing intervention works as intended, the benefit to the child is the same in each of cases A, B, and C (and D and F to boot); namely, the child will be genetically related to both parents and free from a diagnosed condition. If it fails to work as intended, however, the situation is worse in the case of category A than in category C; namely, any resulting child would be left with a more serious and life-limiting condition. These genetic disorders would be diagnosed in the preimplantation embryos, and they would not be transferred. Again, the result is the same in cases A to C: no child. Finally, if the procedure appears to work as intended for the target trait but there is some undetected or unanticipated effect, either because of an unintended alteration or because of pleiotropy, then the risk is, once again, the same in cases A, B, and C: a child affected by iatrogenic effects because of the procedure. In each case there is no more advantage to any resulting child in one category than in the other.

The initiative to lay out a translational pathway for HGE has been provoked, seemingly, less by advances in research than by the announcement that a Chinese researcher, He Jiankui, had transferred edited embryos to women, leading to the birth of the world's first genome-edited babies in China in 2018 (Cyranoski and Ledford 2018). His failure to address an "unmet medical need" (but, instead, to attempt to confer the advantage of HIV resistance on the resulting children) precisely provided the terms in which the anathema, pronounced upon him by his peers at the Second International Summit on Human Genome Editing in Hong Kong in 2018, was framed. Although He's activities were considered naïve, illicit, and downright dangerous from a technical perspective and unethical from the medical perspective, they are, at least, intelligible and coherent from another point of view (which may or may not be correct to impute to him). Rather than selecting as a first case one in which the prospective parents' interests justified the risk of harm, He selected (albeit erroneously and recklessly, as we must not cease to say) a case that may have seemed to him most tractable from a technical point of view and therefore most likely to produce live offspring and least likely to result in iatrogenic harm. Under the medical model, the procedure is undertaken primarily to secure genetic relatedness, and the best outcome is a child unaffected by a genetic disorder. Under what we might think of as He's disruptive "technological model," the principal aim was to demonstrate a potentially game-changing technology while conferring a positive benefit on the resulting child (namely, inherent resistance to an endemic disease). It is a benefit, furthermore, that could not be secured easily by using donated gametes because the relevant mutation is extremely rare in the general population. Regardless of these conjectures, it is reasonable to claim that the origin of the responsible pathway is, temporally and conceptually, at the bifurcation between these different "medical" and "technological" approaches (Baltimore et al. 2018).

The criterion of unmet medical need is one that has been endorsed, implicitly or explicitly, through several reports from learned societies, national academies, and specially convened expert panels. The significance of this approach, however, goes beyond the question of innovation—of securing the *first* use of the technology. If it *were* possible to demonstrate that HGE has a favorable risk profile for a given, contingent outcome, then this would clear the way for HGE to enter the repertoire of alternative treatments for a much wider range of indications, alongside the accepted treatments already in use. Once the innovation hurdle has been surmounted, it might, for example, become the presumptive technology for many cases in which PGD has been hitherto the standard of care. This is not all. Because HGE could, in theory, enable the introduction of novel genetic features unavailable through selective technologies or natural reproduction, it could potentially transform the horizon of possible human agency in reproduction (Nuffield Council on Bioethics [NCOB] 2016). The "most unusual case" could, therefore, offer a bridgehead from which to extend the possible range of applications in other directions—up to a point. The focus on the most unusual case that purports to justify the first use of the new technology obscures a second, perhaps more important effect of the "unmet medical need" claim, namely, to insinuate "medical need" as the presumptive criterion for any eventual use of HGE. It is here that those promoting and opposing HGE appear to be, if not on the same side, at least engaged in the same campaign. Medical need is a criterion that both accept because it offers a bulwark against the insidious expansion of uses of the technology. Neither, therefore, dares question this criterion, only whether and when it may be met.

If Not Unmet Medical Need, then What?

The claim that there is an unmet medical need that could be addressed by HGE is important because it is often the one that supports arguments for HGE research (Deutscher Ethikrat 2019; NAMSRS 2020). The claim that HGE does not address an unmet medical need may be a successful counterargument, but it is not necessarily a successful argument against this innovation. The discussion of unmet medical need has led down a narrow path. What we can now ask is what the argument for HGE might look like if we eschew the medical model that both proponents and opponents of HGE have accepted as the frame in which to represent their positions. Such a task is difficult, but its difficulty illuminates why the assumptions underlying the "unmet medical need" claim have been so hard to relinquish.

I have made the assertion that the aims of HGE are distinctively reproductive rather than therapeutic. Things are, in practice, murkier. Infertility is often a consequence of individual biology (although in the case of HGE the prospective parents may not be technically infertile). Little really hangs on the question of infertility, though. Many would maintain that the anguish that may be experienced by people who are unable to conceive for whatever reason (or to conceive a child who will not perish in infancy or be affected with a lifelong impairment) is comparable to the suffering associated with any form of illness or disability. The important question is not whether this anguish is a *medical* reason but whether it offers a *good* reason for using HGE.

For most potential medical uses of HGE, there appear to be existing alternative technologies, like IVF and PGD, though these treatments are only "alternatives" in some respects. Certainly, these technologies provide a path to parenthood. Using IVF and PGD, the child will be a direct genetic descendant of both parents. Using other technologies, like egg or sperm donation, that child will be genetically related to only one of them or to neither if dual gamete donation or adoption is the chosen path. Moreover, the paths to superficially similar outcomes may themselves be valued very differently: many people might draw a significant distinction between the acceptability of terminating a pregnancy (perhaps several) and of allowing embryos to perish that have only existed in a laboratory. Therefore, if the value placed by the parents on the different procedures and outcomes are allowed to figure into the appraisal, the different approaches are no longer simply "alternatives."

In most societies, a high value is placed on protecting the procreative liberties of people of reproductive age, such that they should not be interfered with for any but the weightiest reasons. Nevertheless, the value, form, and scope of these liberties vary from society to society and from time to time. It may also be the case that individuals' preferences are influenced by prevailing social norms, by the views of others, or by the influence of a powerful industry built on servicing them. This variation implies that what are, in all meaningful senses, real wants and desires may be influenced by contingent moral, epistemological, technological, and economic conditions. Just as HGE is an artifact of human ingenuity, what counts as a "good reason" to use it is not independent of the sociotechnical context.

The discussion so far has been conducted at the individual and intrafamilial levels. As well as affecting the interests of future offspring, the use of reproductive technologies by certain individuals may represent a challenge or threat to other members of society or to the norms of society itself. These externalities may be insidious and difficult to evaluate just because they exist on a different scale. Minimal impact, high prevalence effects, such as stigmatization of parents of children with certain "avoidable" conditions like Down syndrome are notoriously difficult to compare with high impact, low prevalence effects such as the "most unusual cases" described above. Such externalities might include increasing inequality in a way that is corrosive to society, for example because of certain families obtaining access to advantages denied to others or the cumulative social effects of normalizing certain interventions on those in vulnerable positions (NCOB 2018).

It is important to recall that norms are constructed both to support certain reproductive projects and to protect others, especially those in vulnerable positions, from the consequences of the unbridled exercise of reproductive choices. This observation reveals the societal interest in the forms and limits of reproductive assistance available to individuals. A further legitimate, though more abstruse, reason for societal interest arises from the potentially disruptive effect of reproductive technologies on the integrity of the public value system that undergirds a more or less coherent (if dynamic) set of norms that regulate interpersonal relations and conduct. After all, the way societies value reproductive freedoms cannot be distinct from the way they value other kinds of liberties. Norms must negotiate tensions and find consistency with each other while accommodating or resisting technological change.

Some Challenging Implications

Adopting the "medical model" pushed the target for HGE innovation toward the "most unusual case" in a way that made it have little relevance to most people. The question about the circumstances under which it was appropriate to use HGE was thus treated as if it were a continuation of questions about the use of preceding selective embryological technologies. HGE, however, potentially breaks from the former limits of these questions because it can offer genetic relatedness without having to work within the combinatory constraints of the prospective parents' genetic endowment. On one hand, we find that the question of what constitutes a good reason for HGE is freed from the assumption that it must be medical. On the other hand, it potentially opens a wider range of uses and targeted characteristics that would make the efforts and attention given to HGE more explicable than if they were all for the relief of the "most unusual cases" (Baylis and Robert 2005). To accept the criteria of medical use conveniently reduces questions of disposing over the characteristics of future generations to border skirmishes between conservatives and progressives around the limits of legitimate medical practice. Liberating HGE from the medical assumption, however, raises the question of *function creep*, the diversion of the technology to purposes other than those for which it was initially developed. If all uses are arguably "nonmedical" and "nonmedical" uses cannot be ruled out in principle, some other means of moral delineation must be found.

Thinking about HGE outside the "medical model" leads to several potentially discomfiting implications, especially for those involved in HGE research. To do so implicitly challenges the self-image of the researcher as engaged in a project that is about bringing relief to those affected by disease and suffering and translates it to a field in which a range of interests and desires are in tension. Instead of simply treating illness, HGE appears to be to enable choices about what *kind of people* may exist in the future and *whose choices* should hold sway. This is a different and difficult discourse. Researchers and clinicians of good faith are necessarily implicated in addressing these questions, but as *supplicants*, rather than *arbiters*, in a broader moral debate.

There is a further discomfiting implication of unmooring decisions about HGE from the medical model to which they have hitherto been tethered. While the nation state under the rule of law is the natural horizon of public policy and of the authoritative exercise of governance, any policy position about how HGE should be used is contingent on the local interplay between social and technical developments and developments in the global context. On one hand, national policies will come under challenge from pressures within their jurisdiction. On the other, foreign jurisdictions may arrive at contradictory positions that accommodate different ambitions. It is unfortunate that the initiatives of some researchers have turned into threats to others, and, for some, the practice of scientific research has taken on the character of a race or contest (Chen et al. 2020). Not only does public policy need to provide the context for clinical decisions, but states need to recognize their responsibilities to other states as potential resorts for practices that would exert a decadent force on the global moral order. The aforementioned International Commission report dismisses the prospect of an international legal solution to this problem for reasons of sovereignty rather than of subsidiarity (NAMSRS 2020). It remains (at the time of writing) to be seen what recommendations the WHO Expert Advisory Committee on Developing Global Standards for Governance and Oversight of Human Genome Editing will propose in this area.

Before we get carried away, however, it is appropriate to recall that there are many uncertainties about what HGE might achieve, if it achieves anything at all (NCOB 2018). The most likely cases in which HGE could be effective, based on present knowledge, are those of well-characterized traits, conditioned by simple orthogonal, and highly penetrant, genetic factors, such as many of the known single-gene disorders for which the existing alternative reproductive technologies were developed (Horton and Lucassen 2019). Other objectives that cannot be achieved by selective interventions may, nevertheless, be technically tractable for HGE, including some that are not strongly associated with inherited disease (Regalado 2017). Which, if any, of these objectives may be pursued by anyone, in any jurisdiction, is a question in which everyone, everywhere, has a legitimate moral interest.

REFERENCES

Baltimore, David, R. Alta Charo, George Q. Daley, Jennifer A. Doudna, Kazuto Kato, Jin-Soo Kim, Robin Lovell-Badge, et al. 2018. *On Human Genome Editing II: Statement by the Organizing Committee of the Second International Summit on Human Genome Editing.* Washington, DC: National Academies Press. https://www.nationalacademies.org/news/2018/11/statement-by-the-organizing-committee-of-the-second-international-summit-on-human-genome-editing.

Baylis, François, and Jason Scott Robert. 2005. "Radical Rupture: Exploring Biological Sequelae of Volitional Inheritable Genetic Modification." In *The Ethics of Inheritable Genetic Modification: A Dividing Line?* edited by John Rasko, Gabrielle O'Sullivan, and Rachel Ankeny, 131–48. Cambridge: Cambridge University Press.

Chen, Qi, Yonghui Ma, Markus Labude, G. Owen Schaefer, Vicki Xafis, and Peter Mills. 2020. "Making Sense of It All: Ethical Reflections on the Conditions Surrounding the First Genome-Edited Babies." [Version 2; Peer Review: 2 Approved]. *Wellcome Open Res* 5 (216). https://doi.org/10.12688/wellcomeopenres.16295.2.

Cyranoski, David, and Heidi Ledford. 2018. "Genome-Edited Baby Claim Provokes International Outcry." *Nature* 563: 607–08.

Deutscher Ethikrat. 2019. *Intervening in the Human Germline: Opinion (Executive Summary & Recommendations)*. https://www.ethikrat.org/fileadmin/Publikationen /Stellungnahmen/englisch/opinion-intervening-in-the-human-germline-summary .pdf.

Horton, Rachel, and Anneke M. Lucassen. 2019. "The Moral Argument for Heritable Genome Editing Requires an Inappropriately Deterministic View of Genetics." *J Med Ethics* 45 (8): 526–27.

Lander, Eric S., François Baylis, Feng Zhang, Emmanuelle Charpentier, Paul Berg, Catherine Bourgain, Bärbel Friedrich, et al. 2019. "Adopt a Moratorium on Heritable Genome Editing." *Nature* 267: 165–68.

National Academies of Sciences, Engineering, and Medicine (NASEM). 2017. *Human Genome Editing: Science, Ethics, and Governance*. Washington, DC: National Academies Press. https://doi.org/10.17226/24623.

National Academy of Medicine, National Academy of Sciences, and the Royal Society (NAMSRS). 2020. *Heritable Human Genome Editing*. Washington, DC: National Academies Press. https://doi.org/10.17226/25665.

Nuffield Council on Bioethics (NCOB). 2016. *Genome Editing: An Ethical Review*. London: Nuffield Council. http://nuffieldbioethics.org/wp-content/uploads/Genome-editing -an-ethical-review.pdf.

Nuffield Council on Bioethics (NCOB). 2018. *Genome Editing and Human Reproduction: Social and Ethical Issues*. London: Nuffield Council. http://nuffieldbioethics.org/wp -content/uploads/Genome-editing-and-human-reproduction-FINAL-website.pdf.

O'Neill, Onora. 1979. "Begetting, Bearing, and Rearing." In *Having Children: Philosophical and Legal Reflections on Parenthood*, edited by Onora O'Neill and William Ruddick, 25–38. New York: Oxford University Press.

Parfit, Derek. 1984. *Reasons and Persons*. Oxford: Clarendon Press.

Qasim, Waseem, Persis Jal Amrolia, Sujith Samarasinghe, Sara Ghorashian, Hong Zhan, Sian Stafford, Katie Butler, et al. 2015. "First Clinical Application of Talen Engineered Universal CAR19 T Cells in B-ALL." *Blood* 126 (23): 2046.

Regalado, Antonio. 2017. "Engineering the Perfect Astronaut." *MIT Technology Review*, April 15, 2017. https://www.technologyreview.com/2017/04/15/152545/engineering -the-perfect-astronaut/.

13

Genome Editing, in Time

Robert Sparrow

New technologies come to us from the future. We dream of them in science fiction: these dreams shape our goals as we work to make them a reality. We imagine new technologies, though, as if they have just been developed but are also, implausibly, fully realized. We seldom imagine them as old news. Rather, we conceive of our technological future primarily as a present yet to arrive.

Thinking about new technologies on the model of a future *present* risks obscuring an important set of social and ethical issues that relate to the development of new technologies and the aging of existing technologies. Once technologies are realized, they age and become old technologies: they are often old for longer than they were new. The generation making decisions about the shape of the technology is often not the generation that will have to live with the consequences of their decisions. To carefully consider the wisdom of developing new technologies, we must do more than think about them in a timely fashion—we must think about them "in time." We must understand them as projects that have a temporal dimension, such that the ethical issues they raise alter as they move from the future to the present and then into the past.

What lessons can we learn to guide us now, and in the years to come, by thinking about the temporality of heritable genome editing (HGE)? The ethics of HGE, once we know how to do it safely, is quite different from the ethics of HGE while it remains an experimental technology with unknown risks. Moreover, many of the ethical issues that will arise once HGE is safe and effective have barely been discussed because they involve a difficult-to-imagine future in which HGE is old news. The need to think about this technology in time is critical because germline interventions will affect all the descendants of modified individuals.

The Real Use-Case for HGE Is Not Therapy but Enhancement

In developing new technologies, scientists and engineers are shaping the future. The more powerful a technology is advertised as being, the more the public should be consulted about whether it should be developed, and the more input the public should have in determining its form (Sparrow 2008).

The prospects for meaningful public input into debates about HGE at the current moment are especially vexed because the story the public is being told about the importance of developing this technology verges on dishonest. Public discussions of genome editing are focused on its potential to help people with family histories of genetic disease to have "healthy babies." However, an existing technology, preimplantation genetic diagnosis (PGD), allows prospective parents to achieve this end in most cases. Only in exceedingly rare cases, where a couple is unable to produce an embryo without genes associated with an impairment (for example, where both parents have homozygous deafness), might genome editing have something to contribute here. Moreover, even in such cases, prospective parents might achieve the birth of a child without genes associated with impairments by using donor gametes. The use of HGE cannot therefore be justified with reference to the health of the future child. If there is a justification for the use of genome editing here, it must refer to the moral weight of the prospective parents' desire to have a genetically related child without genes associated with impairments (Baylis 2019). The therapeutic justification for the development of HGE is correspondingly weak (Darnovsky and Hasson 2020).

A stronger case can be made for using HGE to *enhance* future human beings. Genome-editing technologies like CRISPR make it possible to add new genes to an organism's genome or tweak the function of existing genes to increase the organism's capacities beyond what is species typical. The potential of CRISPR/*Cas9* and other such tools to enhance the capacities of future individuals is arguably the real reason researchers are keen to pursue them (Regalado 2015). Indeed, it is striking that the only use of HGE to bring children into existence to date was intended to provide an enhancement in the form of increased immunity to HIV (Kuersten and Wexler 2019). Once a technology of genetic human enhancement exists, the social and economic pressures to use it to provide one's children with a competitive advantage are likely to be overwhelming. Thus, the world that researchers are likely to bring about is not the one they advertise.

Reconciling with Risk to Realize the Future

For the moment, a safe and effective technique of HGE remains firmly in the future. The *ethical* issue that stands between now and a future where HGE is possible, which is receiving the most attention within the scientific community, is how to develop and test genome editing without subjecting edited individuals—and the women who will give birth to them—to unconscionable levels of risk (National Academies of Sciences, Engineering, and Medicine [NASEM] 2020; WHO Expert Advisory Committee on Developing Global Standards for Governance and Oversight of Human Genome Editing 2021). Experiments on mice and nonhuman primates will only take us so far. Eventually, the technology will have to be tested in humans. How can we ethically show that HGE is safe and effective when its first uses in humans will necessarily be experimental and the persons most affected—the edited individuals and their descendants—cannot give their consent to the procedure?

Thinking about HGE in time highlights the true extent of this challenge. If one really wants to show that HGE is safe, it will not be enough to show that the procedure succeeds in modifying only the targeted genetic sequences and that genetically modified children are born healthy. Genes interact with environments to produce effects on the phenotype of the organism across the entirety of the life course. Thus, it is possible to conclude that modified individuals are no worse off than unmodified individuals only once enough modified individuals have lived out the entirety of their natural lifespans. If we are serious about wanting to ensure that this technology is "safe," we will need to create and maintain institutions that allow us to track the health and well-being of those who have been edited across their entire lives. Ideally, we would be able to identify those who carried out any given set of edits and hold them responsible should the edits not perform as advertised.

The length of time before genetic edits reveal their effects in human beings also represents a profound problem for attempts to *develop* human enhancements. For many modifications, we will not know whether they improve a modified individual's capacities until they have grown up. Thus, the time required to test each generation of genome-editing technology is likely to be at least 15–20 years. Even then, we will not know whether the new genes will improve the lives of "enhanced" individuals overall. Given that we should anticipate that new enhancements will continue to be proposed, attempts at human enhancement using this technology are likely to remain experimental for many years, even if the basic technology of genome editing proves to be safe.

There is a way to overcome the concerns about risk to the edited individuals, toward which some advocates of the development of genome editing occasionally gesture. Like many reproductive technologies, HGE would require parents to conceive via in vitro fertilization (IVF). If parents choose to undergo IVF to pursue HGE then, by changing the time of conception, this technology will lead to the birth of a different child than the one they might have conceived naturally (Sparrow 2021). In the terms popularized by Derek Parfit (1984, 351–79), in practice, germline editing would almost always be "identity affecting" rather than "person affecting": it will change who will be born rather than alter the characteristics of a particular embryo. Thus, the counterfactual that we use to assess whether our actions will harm or benefit others ("what will the individual's welfare be like if we do not act as we intend?") fails: were prospective parents to choose not to use genome editing, a different child would be born—and it is difficult, if not impossible, to know how to compare the relative merits of existence and nonexistence.

The ethics of identity-affecting interventions is highly contested. However, a plausible position insists that, except in the rare cases where existence is worse than nonexistence, individuals are not harmed by being brought into existence (Brock 1995). According to one line of thinking, then, all that is required for it to be ethical to genome edit human beings is that there is little risk that future individuals' lives will be so wretched that they wish that they had never been born (Feinberg 1986). This is a much easier standard to meet!

If we are to rely upon the claim that HGE is not person affecting to justify proceeding to first-in-human trials, though, the arguments about safety and risk that currently dominate public deliberations about genome editing are distractions. Moreover, there is good evidence from the history of the development of other reproductive technologies that this is the case. We have only a little more than forty years of evidence about the impacts of being conceived via IVF on individuals' health, yet this technology has been used to conceive more than eight million children (European Society of Human Reproduction and Embryology 2018). The impacts of PGD, and of intracytoplasmic sperm injection, over the full human life-course are similarly unknown. Mitochondrial replacement therapy is currently being trialed in human beings with no indication that researchers are willing to wait eighty-odd years before declaring it to be safe. The fact that previous generations of reproductive technologies were made available to the public before their consequences over the full course of a human life were known strongly suggests that concerns about safety are unlikely to stand in the way of HGE being used.

Aging Technologies Generate New Ethical Issues

The issues discussed above are, of course, only some of the ethical issues that we must confront as we decide whether—and how—to pursue heritable genome editing: they are the issues associated with the futurity of this technology that are most illuminated by thinking about the importance of time. Should safe and effective HGE ever become possible, another set of ethical questions, associated with the movement of this technology in time from new to old, will arise. Once-new technologies live on, sometimes in out-of-the-way places, sometimes everywhere, for much longer than we think (Edgerton 2008).

The fate of those who are enhanced, as well as the social impacts of the availability of enhancement, will be shaped by the fact that, in all likelihood, genome-editing technology will constantly be improving. As companies compete to develop and market this technology, we should expect the rate of the release of improved enhancements to accelerate. The technology available to those who wish to genetically modify their embryos at any date will not be as powerful as the technology that would be available to them in another ten years.

This has several implications (Sparrow 2015, 2019). Most immediately, rapid progress in the development of HGE will mean that prospective parents considering editing the genomes of their future children will face a dilemma familiar to many of us when we confront decisions about purchasing consumer electronics: no matter when we decide to purchase a new technology, a better version will shortly become available, rendering the one we have purchased obsolete. Because the longer one delays having children, the better their enhancements will be, it will be extremely difficult to commit to having children at any point (Sparrow 2019). Parental "option regret" is also likely to increase because of this dynamic: "If only we had waited, our child would have been so much better."

The fact that genome editing will render human beings subject to the same sort of dynamics that affect products and technologies lends weight to arguments advanced by Hans Jonas (1974) and Jürgen Habermas (2003), among others, that it will blur the distinction between "the born" and "the made." In the future, our nature—or at least our genes—will reflect the choices of others. This, in turn, may affect our ability to understand ourselves as the sole authors of our lives.

Eventually, continuous development of HGE technology will result in a new form of genetic, and perhaps social, stratification according to date of conception, as each generation receives better enhancements than the one before. It may also lead to what were once "enhanced" genomes becoming out of date—even "obsolete," in the sense that they no longer generate the benefits they were intended to secure. While (some) people born in any particular year will have superior enhancements to older individuals, younger individuals will have even better enhancements. Although any absolute benefits associated with the enhancements will not be affected by this, it will jeopardize any benefits associated with being able to perform better, in some regard, than those around one. What were once enhancements will eventually end up as disadvantageous. This is especially worrisome because people will spend more of their lives with obsolete than state-of-the-art enhancements.

These concerns may seem like science fiction now. It may be that genome editing never becomes more than a minor footnote in the list of humanity's technological fantasies. The time it takes to show that any edit achieves what it is advertised to do may mean that the development of this technology proceeds so slowly that concerns about obsolescence or a genetic "rat race" (Sparrow 2015) never arise. Nevertheless, if we want to deliberate carefully—and choose wisely—about new technologies, we must think about them "in time." Both the role played by time in the movement of genome editing from promise to reality and the issues that will arise as HGE ages deserve more attention than they have yet received.

Acknowledgments

This research was supported by the Australian Government through the Australian Research Council's Discovery Projects scheme (project DP170100919). The views expressed herein are those of the author and are not necessarily those of the Australian Government or Australian Research Council.

REFERENCES

Baylis, Françoise. 2019. *Altered Inheritance: CRISPR and the Ethics of Human Genome Editing*. Cambridge: Harvard University Press.

Brock, Dan. 1995. "The Non-Identity Problem and Genetic Harms—The Case of Wrongful Handicaps." *Bioethics* 9 (3/4): 269–75.

Darnovsky, Marcy, and Katie Hasson. 2020. "CRISPR's Twisted Tales: Clarifying Misconceptions about Heritable Genome Editing." *Perspect Biol Med* 63 (1): 155–76.

Edgerton, David. 2008. *The Shock of the Old: Technology and Global History Since 1900*. London: Profile Books.

European Society of Human Reproduction and Embryology. 2018. "More than 8 Million Babies Born from IVF since the World's First in 1978." *ScienceDaily*, July 3, 2018. https://www.sciencedaily.com/releases/2018/07/180703084127.htm.

Feinberg, Joel. 1986. "Wrongful Life and the Counterfactual Element in Harming." *Soc Philos Policy* 4 (1): 145–78.

Habermas, Jürgen. 2003. *The Future of Human Nature*. Cambridge, England: Polity Press.

Jonas, Hans. 1974. "Biological Engineering–A Preview." In *Philosophical Essays: From Ancient Creed to Technological Man*, edited by Hans Jonas. Englewood Cliffs, NJ: Prentice Hall.

Kuersten, Andreas, and Anna Wexler. 2019. "Ten Ways in Which He Jiankui Violated Ethics." *Nat Biotechnol* 37 (1): 19–20.

National Academies of Sciences, Engineering, and Medicine (NASEM). 2020. *Heritable Human Genome Editing*. Washington, DC: National Academies Press. https://doi.org/10.17226/25665.

Parfit, Derek. 1984. *Reasons and Persons*. Oxford: Clarendon Press.

Regalado, Antonio. 2015. "Engineering the Perfect Baby." *MIT Technology Review*, March 5, 2015. https://www.technologyreview.com/2015/03/05/249167/engineering-the-perfect-baby/.

Sparrow, Robert. 2008. "Talkin' 'bout a (Nanotechnological) Revolution." *IEEE Technol Soc* 27 (2): 37–43.

Sparrow, Robert. 2015. "Enhancement and Obsolescence: Avoiding an 'Enhanced Rat Race.'" *Kennedy Inst Ethics J* 25 (3): 231–60.

Sparrow, Robert. 2019. "Yesterday's Child: How Gene Editing for Enhancement Will Produce Obsolescence—And Why It Matters." *Am J Bioeth* 19 (7): 6–15.

Sparrow, Robert. 2021. "Human Germline Genome Editing: On the Nature of Our Reasons to Genome Edit." *Am J Bioeth* 22 (9): 4–15.

WHO Expert Advisory Committee on Developing Global Standards for Governance and Oversight of Human Genome Editing. 2021 *Human Genome Editing: A Framework for Governance*. Geneva: World Health Organization.

VI OVERSIGHT AND MONITORING

Regulating CRISPR

A Quest to Foster Safe, Ethical, and Equitable Innovation

Andrew C. Heinrich

While the scientific community was celebrating Emmanuelle Charpentier and Jennifer Doudna's Nobel Prize for their pioneering work on CRISPR in 2020, Victoria Gray was celebrating a milestone of her own: one year symptom-free after a lifetime of battling sickle cell anemia. Ms. Gray is the first patient successfully treated with a stem cell therapy using stem cells edited with CRISPR-*Cas9* technology (Kaiser 2020). In the same year, halfway around the world in Uganda, scientists began testing genetically modified cassava, a staple crop providing somewhere between a third and a half of calories consumed in sub-Saharan Africa in the hope of finding a way to mitigate the devastating impact of cassava brown streak disease (Agaba 2020). Meanwhile, in the United States, scientists published findings of the first successful use of CRISPR technology on mitochondria (*Nature* 2020).

In approximately two decades, CRISPR technology has evolved at a rapid pace from an experimental technology used in the lab to make corrections to single point mutations to a widely implemented technology leading to breakthroughs in human, animal, and plant health. Further, the vast future potential of CRISPR is shown by its many unrealized theoretical uses, such as HIV therapy (Dash 2019). Many have argued that CRISPR could be central to numerous improvements in lives and livelihoods around the world, from addressing malnutrition in the developing world to eradicating crippling genetic diseases in humans and even avoiding extinction of certain animal species.

Despite CRISPR's enormous potential, it has also faced significant criticism. First, there are profound ethical concerns about parents genetically engineering their ideal traits in their child. The idea of a child being genetically enhanced—rather than genetically modified to avoid a genetic disease like sickle cell anemia—with the intention to reflect the physical ideals of the child's parents is horrifying to many, including even some of the most ardent CRISPR advocates. Second, any use of CRISPR on germline cells raises necessary questions regarding the future child's inability to consent. The unease around consent in the near term is compounded by the fact that risks of germline editing are not yet fully known. Third, safety itself is a concern. Off-target edits may cause unforeseen, unwanted consequences that can be damaging. CRISPR also carries the risk of mosaicism, a condition which causes an organism to carry more than one genetic line, resulting in a wide variety of symptoms ranging from skin problems to intellectual disability. Finally, there are profound equity concerns, both of access and research focus: CRISPR should be used to address the most pressing needs of our time, and research should prioritize mechanisms by which to do so.

Regulation of CRISPR technology is still in its infancy. As with any emerging technology, regulators are in an unenviable position of needing to make complex decisions and create multifaceted regulatory regimes quickly. Any CRISPR regulatory regime would need to foster innovation and widespread use, especially use that could affect the developing world and underserved populations, while also guarding against ethical and safety concerns. The age-old questions in regulating a new technology are those of fit. Is there an existing agency or framework that suits the new technology? Is there a combination thereof that might collectively serve the purpose? Or is an entirely new apparatus required? The questions are particularly complex in the case of CRISPR, because it is a technology with uses in two distinct domains: human health and agriculture. Should all CRISPR be regulated together, or should the distinct uses be regulated separately?

The History of CRISPR Regulation

Since the first description of a CRISPR locus in 1993, scientific understanding and technological use of CRISPR technology has evolved rapidly. The discovery of *Cas9* in 2005 sparked a series of innovations that eventually led to CRISPR-*Cas9* gene-editing technology in 2013 (Broad Institute 2022), and development of the technology and its applications has moved at a breakneck speed for the past decade.

Lawmakers and regulators have been in a doomed race against CRISPR-*Cas9* innovation. In the United States, the first regulatory and legislative action was one of hedged inaction. In 2016, Congress passed the National Bioengineered Food Disclosure Standard, which directed the USDA to create a national standard for mandatory labeling of genetically modified foods (USDA 2018). Though CRISPR-*Cas9* technology was already being used on seeds and in some early animal gene-editing trials, the bill was agnostic on whether CRISPR-modified foods would have to be labeled as bioengineered. The USDA's Agricultural Marketing Service (AMS) guidance based on the Food Disclosure Standard stated that CRISPR-modified foods would be treated on a case-by-case basis without providing any evaluation framework, only to drop all such regulations in April 2021 (USDA 2018). The only subsequent federal legislation that addressed CRISPR in any way was the *Bioeconomy Research and Development Act* of 2020, which provides funds for initial research into the cybersecurity consequences of storing certain genetic information, including CRISPR sequences (S.3734 2019–20). The US Supreme Court's only treatment of CRISPR-related issues has ruled on matters of intellectual property rights, finding that a patent cannot be issued for a naturally occurring genetic sequence but can be issued for a product or technology (*Association for Molecular Pathology v. Myriad Genetics, Inc.* 569 U.S. 576, 2013).

There are two notable trends in CRISPR regulation in the United States to date. The first is the relative lack of action. There is no shortage of questions around CRISPR to be addressed, yet only two passed and ratified federal statutes empower agencies to take action related to CRISPR, and the resulting agency action has been noncommittal. The second and far more striking trend is that the little legislative and regulatory action that has been taken does not address any of the central ethical or safety issues that constitute a majority of scientific and public concern about CRISPR technology.

Though none of the branches of the federal government has taken decisive action on the safety and ethical dimensions of CRISPR, faulting them for inaction might be unfair. Most importantly, many of the primary concerns around safety and ethics cannot be addressed well without further research. In addition, much of the technical expertise in government comes from industry professionals transitioning into policy work. Because CRISPR is under a decade old, there simply has not been enough time for the government to gain CRISPR experts.

The first United States entity to directly address CRISPR through legislation is the state of California. *California Senate Bill 180*, signed into law in 2019, requires that do-it-yourself gene therapy kits using CRISPR technology be accompanied by a warning label when marketed or sold in the state of California. S 180 explicitly cites safety concerns, especially when individuals attempt to self-administer CRISPR therapies. The bill falls short of placing any restrictions on the sale or use of the kits, noting that the rapid evolution of CRISPR technologies means that more research is required to determine their safety before any more definitive action can be taken. Even if more research were available, it is questionable that the California state legislature could have constitutionally restricted the sale of the kits in California, because the federal government is responsible for regulating the approval of medical products for sale in the United States. S 180 purports to be the first piece of legislation to directly address the use of CRISPR technology anywhere in the United States (although the FDA states in nonlegally binding documentation that the sale of do-it-yourself CRISPR kits is illegal, there is no explicit federal law on the matter), and it might speak some truth to Brandeis's claim that states are laboratories for innovation (*New State Ice Co. v. Liebmann*, 285 U.S. 262, 1932). However, as the preamble to S 180 itself recognizes, federal legislation and regulation is necessary for two reasons. First, the regulation of the technology itself is a federal matter, so to go beyond regulation of marketing would necessarily require federal action. Second, even with respect to marketing, a patchwork of inevitably inconsistent state regulations would be a suboptimal outcome.

CRISPR regulation outside of the United States has been similarly sparse and noncommittal. Some jurisdictions and international organizations have hosted conferences and debates, but there is still a dearth of meaningful legislation that addresses the core CRISPR issues. No doubt there is reason to wait: more research is needed to identify the nature and degree of the risks, more debate is required around the ethical implications, and all the while, CRISPR technology will continue to evolve at a rapid pace. The European Union provides a powerful cautionary tale about the perils of failing to regulate CRISPR quickly and meaningfully as use of the technology grows more prevalent. In July 2018, the Court of Justice of the European Union (CJEU) handed down its ruling in C-528/16, in which it found that animals and plants on which CRISPR therapies had been used were Genetically Modified Organisms (GMOs) and subject to the EU's stringent Directive on GMOs (CJEU 2018). Until the CJEU's determination, it was understood that the EU had chosen to define GMOs only as organisms treated with transgenesis, which involves the introduction of foreign genetic material (Ledford 2019). It was without any legislative or regulatory guidance that the CJEU decided that those obtained through mutagenesis, or the altering of genes in the germline that forms the organisms, were also GMOs. This surprise ruling caused pandemonium in the European agriculture industry, prompting the European Commission to form a working group and report on the need for coherent policy to ensure that scientific expertise drives decision-making with respect to which food products can be marketed and sold in Europe, rather than a blanket judicial decision (van der Meer 2021). The EU's experience demonstrates the importance of prompt, deliberate action that directly addresses the regulatory needs surrounding CRISPR technology.

Relevant Considerations in CRISPR Regulation

Any full regulatory framework in using CRISPR should address the following components:

1. **Safety Considerations**: Though more work is to be done, there is a preliminary awareness of certain risks inherent in using CRISPR technologies on human, animal, and plant subjects. Regulation can help guard against those risks and promote research to better understand them. As our understanding grows, regulation can be revised accordingly. Commonly used mechanisms to guard against safety concerns in other regulatory contexts include requiring training and licensure to use the technology, restricting the facilities in which the technology can be used, restricting use to certain contexts in which the need outweighs the risk, requiring labeling and other notifications of risks, requiring a higher threshold of informed consent, creating a licensing process for companies wishing to engage in business relating to novel CRISPR technologies, and requiring injury compensation programs.

2. **Ethical Considerations**: Arguably the most important ethical consideration, brought to the fore by the "CRISPR baby" scandal, is whether heritable genome editing is permissible. Chief among concerns in the popular press is that using this technology will amount to a eugenics program by way of genetic modification, but there are also concerns about therapeutic legitimacy, risk and uncertainty, and responsibility to future generations (Schleidgen et al. 2020). In any genome editing performed on zygotes, the ethical definition of informed consent when risks are still underresearched is unclear. Many prominent scientists have called for a moratorium on heritable genome editing (Lander et al. 2019).

   Regulation must address these justifiable concerns by ensuring that uses of CRISPR technology on human germlines are restricted to certain purposes. There are few instances where the precautionary principle is so justified, given the unknown, intergenerational risks. As the knowledge base improves over time, perhaps there will be grounds to gradually permit certain, limited uses of CRISPR technology on human germlines according to the latest scientific knowledge, including perhaps—if there is a desire to leave more autonomy with patients and their care providers—by providing a balancing test to guide care providers in determining appropriate use. There are also inevitable ethical considerations related to research on human germline cells and to the ethical definition of informed consent when risks are still underresearched. Unforeseen consequences, like off-target mutations, will not only affect the child, but also their progeny potentially for generations. Because the person who will live with the consequences of these potential risks inherently cannot consent, adhering to the precautionary principle pending further research might be particularly wise in this case.

3. **Encouraging Research and Innovation:** Any regulatory regime must encourage research and innovation. While some fear that a CRISPR regulatory regime will stifle innovation, it can actually help to support it. Regulatory regimes often motivate the creation of directed funding opportunities prioritizing areas of innovation that may improve quality of life, especially those that might not be the most inherently profitable.

4. **Promoting Justice and Equity:** Finally, but of equal importance, is the need to promote justice and equity. Addressing this concern begins with clinical trials, especially given the horrific history of racial injustice in many clinical trials. Considerations of justice and equity then proceed to research considerations: it is of the utmost importance that research agendas and funding streams prioritize the uses of CRISPR that are most helpful to the underserved, both in the United States and globally. Finally, it is critical that underserved populations can access this new technology. Otherwise, development will simply perpetuate inequality.

Toward a CRISPR Regulatory Regime in the United States

An age-old question in US administrative law is whether existing agencies can adequately address a novel regulatory challenge or whether a new agency is needed. Any answer to this question rests on several factors: the components and nature of the underlying technology itself, its producers and the production process, its uses, whether it is intended to be used by trained professionals or the public, and its risks. CRISPR can be categorized in many ways with respect to these criteria, but at its core, it has two regulatory identities: a biomedical technology that interfaces with human health and a technology that interfaces with our food sources. CRISPR can therefore be regulated by existing agencies, albeit with new infrastructure, and perhaps new statutory authority.

CRISPR, as a human health biotechnology, can be adequately regulated by the FDA, with the addition of regulatory authority for CRISPR for cosmetic use; and CRISPR as a food-oriented biotechnology can be adequately regulated by the USDA. Because the FDA considers CRISPR to be a gene-therapy product when used on humans, innovations can be adequately evaluated via the Center for Biologics Evaluation and Research (CBER) before they come to market. Two CRISPR concerns are unaddressed by the traditional FDA approval process, however. The first is mitigating new safety risks as they become known when an approved technology is used by a larger patient population. Fortunately, this concern is being addressed by existing postmarket regulation modalities, also known as *Phase 4 Regulation* (FDA 2022). This is an increasingly used tool throughout the FDA and expanding its scope has been discussed (Brower 2007). Especially in the early days of CRISPR, the use of Phase 4 Regulation will be vital in gaining a better understanding of safety and efficacy as CRISPR technology is used to treat a larger and more heterogeneous population. Phase 4 Regulation is already being used similarly in other cases of novel treatments; therefore, the FDA's work in this area will be simply a continuation of existing trends in FDA regulation (Lee et al 2020).

The second concern, however, exposes an area in which existing FDA regulatory authority may be insufficient: addressing some of the notable ethics and equity concerns raised by CRISPR. For example, ethical concerns around using CRISPR to influence phenotypes for cosmetic rather than health purposes can be addressed only by federal legislation that gives the FDA authority to regulate based on cosmetic use. This legislation could take numerous forms and could impose restrictions (or not) on how one accesses the technology for cosmetic aims, who can administer it, and for what purpose. The fact that this legislation does not yet exist does not mean that the FDA would be ill-equipped to handle these statutory authorities if they were given to it, however. Indeed, quite to the contrary, the FDA is well-equipped to carry out whatever mandate it might be given by Congress in restricting use of CRISPR based on ethical grounds.

Some may argue that these layers of additional FDA regulation could stifle research and innovation. Fortunately, two solutions can help circumvent this risk. First, the FDA can incentivize targeted innovation in specifically high need areas. For example, using orphan drug status, innovators could be incentivized to develop CRISPR-based products that address genetic diseases affecting particularly low-income or small populations. Second, outside of the FDA, NIH funding can extensively encourage better understanding of ethical and safe CRISPR use. In fact, these approval processes and funding streams for CRISPR research can incentivize more focused innovation in areas that need it most. Building on the theory of technology-forcing regulation, these focused funding streams could benefit CRISPR and its uses: the regulations put in place, rather than acting as a hindrance, serve to force technological innovation in areas deemed to be most desirable, safe, and effective. Indeed, decades from now, the idea that CRISPR regulation might have stifled innovation could seem as preposterous as claims made a century ago that basic clinical trial ethics were going to stifle innovation.

The USDA is already well-equipped to regulate CRISPR as a food-oriented biotechnology. It has already been instructed to regulate labeling requirements for food using CRISPR-modified organisms (USDA 2018). As science and congressional mandates progress over time, the USDA will continue to respond to the regulatory needs for CRISPR's uses on animals and plants. The USDA's CRISPR regulatory framework might be simpler, because it can be reduced to a matter of safety without the same profound ethical challenges that the FDA must confront with respect to use on human germlines.

One area where the USDA could be particularly helpful with respect to innovation for the purposes of justice and equity is through its Foreign Agricultural Service (FAS), a network of USDA offices in the United States and abroad that aim to provide food exports and technical assistance internationally. Given CRISPR's outstanding potential to combat food shortages in the developing world, FAS could play a crucial role in exporting CRISPR-modified food sources, seeds, and livestock to areas where they are most needed. The FAS could also play a role in advancing research related to agricultural uses of CRISPR in parts of the world where it is needed most, such as through its existing Scientific Cooperation Research Program.

CRISPR Regulation in Africa

Though CRISPR is often discussed in terms of its current and potential human uses, some of its greatest impacts on humankind might come through its uses for plants and animals. Approximately 10% of the world is undernourished, and CRISPR has the potential to create more resilient and productive plants and livestock that can improve access to food in parts of the world where it is needed most (Graziano da Silva et al. 2019). The African continent has already realized the promise of CRISPR and has begun to experiment with ways it can create disease-resistant crops and more resilient livestock. Scientists across Africa have been making inspiring progress in applying CRISPR technology to several agricultural solutions.

The African Medicines Agency (AMA), which will coordinate medical regulation throughout Africa, could promote and oversee a continent-wide discussion of regulating CRISPR technology across the same two dimensions proposed for the United States: human uses and animal and plant uses (Anderson 2017). The potential for CRISPR in both domains is particularly extraordinary in Africa. In the human use domain, the future AMA could incentivize research on applying CRISPR technologies to genetic diseases most prevalent in Africa. Likewise, with respect to animal and plant uses, a similar continent-wide authority could ensure that crops like cassava, and others most essential for their nutritional value, are systematically researched with the greatest focus and investment.

Regulatory harmonization would be a welcomed step toward incentivizing coordinated and directed efforts in CRISPR research and use. The East African Community is an early standard-bearer for regulatory harmonization in Africa and has demonstrated the benefits of harmonization for streamlining industry innovation and growth (Mashingia et al. 2020). Continent-wide harmonization on CRISPR, with respect to human use in the future AMA and to animal and plant use, would bring with it hope for significant improvement to quality of life for Africans across the continent.

Conclusion

Designing a regulatory regime for CRISPR is necessary and timely. Though it is immeasurably challenging to regulate a constantly evolving, moving target, it is precisely because of CRISPR's rapid technological development that regulation is needed. The EU's cautionary tale is one to be heeded: deliberate regulation, even if imperfect and intended to evolve alongside the science, is crucial to avoid any surprises from other sources, such as the judiciary. Next, regulating CRISPR requires not one but two regulatory regimes: one for its uses as a human gene therapy, and another for its uses as a plant and animal gene therapy. In the United States, the FDA is sufficiently equipped to regulate human uses, especially safety concerns, but needs new, specific congressional authorization to regulate CRISPR across ethical, justice, and equity dimensions and more specific guidance on the dimension of fostering innovation. The USDA is currently even better equipped to regulate animal and plant uses, because the concerns are far more one-dimensional and solely a matter of safety. Despite the straightforward nature of CRISPR's use on animals and plants, Congress may choose to provide guidance to the USDA. Crucially, and perhaps overlooked, the FSA has the potential to give the USDA a vehicle for contributing to the justice and equity dimension of CRISPR regulation by ensuring that research, the technology itself, and CRISPR-enhanced agricultural products are available in parts of the world where they are needed most. United States regulatory agencies are ready for the ongoing challenges of regulating CRISPR; they only need congressional authority and direction to do so in ways that respond to both the opportunities and concerns CRISPR raises.

In Africa, CRISPR has a unique opportunity to contribute to nutrition, health, and social mobility. Regulatory harmonization across the continent regarding human and agricultural uses of CRISPR could result in monumental progress on both fronts.

Finally, regulation by international organizations, such as the World Health Organization, is always an open question. Eventually, this international regulation may be desirable or necessary, but it would also create challenges, especially on the human use front, given differing international views. It might be wise for international organizations to continue serving as facilitators, including hosting conferences or funding research, to further explore their potential roles.

REFERENCES

Agaba, John. 2020. "Ugandan Scientists Use CRISPR in Pioneering Research to Breed Hardier Cassava." *Cornell Alliance for Science*, April 29, 2020. https://allianceforscience.cornell.edu/blog/2020/04/ugandan-scientists-use-crispr-in-pioneering-research-to-breed-hardier-cassava/.

Anderson, Tatum. 2017. "Enter the African Medicines Agency, Continent's First Super-Regulator?" *International Policy Watch*, March 7, 2017. https://www.ip-watch.org/2017/07/03/enter-african-medicines-agency-continents-first-super-regulator/.

Association for Molecular Pathology v. Myriad Genetics, Inc, 569 U.S. 576 (2013).

Broad Institute. 2022. "CRISPR Timeline." https://www.broadinstitute.org/what-broad /areas-focus/project-spotlight/crispr-timeline.

Brower, Amanda. 2007. "Phase 4 Research Grows Despite Lack of FDA Oversight." *Biotechnol Healthc.* 4 (5): 16–22.

California Congress. Senate. *Gene Therapy Kits: Advisory Notice and Labels, 2019–2020.* S 180. 2019–20 Cong. https://leginfo.legislature.ca.gov/faces/billTextClient. xhtml?bill_id=201920200SB180.

CJEU C-528/16 (2018).

Dash, Prasanta K., Rafal Kaminski, Ramona Bella, Hang Su, Saumi Mathews, Taha M. Ahooyi, Chen Chen, et al. 2019. "Sequential LASER ART and CRISPR Treatments Eliminate HIV-1 in a Subset of Infected Humanized Mice." *Nat Commun* 10 (2753). https://doi.org/10.1038/s41467-019-10366-y.

"(FDA) Food and Drug Administration Department of Health and Human Services, Subchapter D—Drugs for Human Use." *Code of Federal Regulations*, title 21 (2022): Vol. 5. https://www.accessdata.fda.gov/scripts/cdrh/cfdocs/cfcfr/CFRSearch .cfm?fr=312.85.

Graziano da Silva, José Golbert F. Houngbo, Henrietta H. Fore, David Beasley, and Tedros Adhanom Ghebreyesus. 2019. *World hunger Is Still Not Going Down after Three Years and Obesity Is Still Growing—UN Report.* https://www.who.int/news/item/15-07-2019 -world-hunger-is-still-not-going-down-after-three-years-and-obesity-is-still-growing -un-report.

Kaiser, Jocelyn. 2020. "CRISPR and Another Genetic Strategy Fix Cell Defects in Two Common Blood Disorders." *Science*, December 6, 2020. https://www.sciencemag.org /news/2020/12/crispr-and-another-genetic-strategy-fix-cell-defects-two-common -blood-disorders.

Lander, Eric S., Françoise Baylis, Feng Zhang, Emmanuelle Charpentier, Paul Berg, Catherine Bourgain, Bärbel Friedrich, et al. 2019. "Adopt a Moratorium on Heritable Genome Editing." *Nature* 567. https://doi.org/10.1038/d41586-019-00726-5.

Ledford, Heidi. 2019. "CRISPR Conundrum: Strict European Court Ruling Leaves Food-Testing Labs without a Plan." *Nature* 572 (15). https://doi.org/10.1038/d41586-019 -02162-x.

Lee Grace, José Romero, and Beth P. Bell. 2020. "Postapproval Vaccine Safety Surveillance for COVID-19 Vaccines in the US." *JAMA* 324 (19): 1937–38. https://doi.org/10.1001 /jama.2020.19692.

Mashingia, Jane H., Vincent Ahonkhai, Noel Aineplan, Aggrey Ambali, Apollo Angole, Mawien Ari, Samvel Azatyan, et al. 2020. "Eight Years of the East African Community Medicines Regulatory Harmonization Initiative: Implementation, Progress, and Lessons Learned." *PLoS Med* 17 (8). https://doi.org/10.1371/journal.pmed.1003134.

Nature. 2020. "Mitochondrial Genome Editing: Another Win for Curiosity-Driven Research." 583 (332). https://doi.org/10.1038/d41586-020-02094-x.

New State Ice Co. v. Liebmann, 285 U.S. 262 (1932).

Schleidgen, Sebastian, Hans-Georg Dederer, Susan Sgodda, Stefan Cravcisin, Luca Lüneburg, Tobias Cantz, and Thomas Heinemann. 2020. "Human Germline Editing in the Era of CRISPR-Cas: Risk and Uncertainty, Inter-Generational Responsibility, Therapeutic Legitimacy." *BMC Medical Ethics* 21. https://doi.org/10.1186/s12910-020-00487-1.

U.S. Congress. Senate. *Bioeconomy Research and Development Act of 2020*. S 3734. 2019–20
    Cong., 2nd sess. https://www.congress.gov/bill/116th-congress/senate-bill/3734.
(USDA) U.S. Department of Agriculture. 2018. *Establishing the National Bioengineered
    Food Disclosure Standard*. https://www.usda.gov/media/press-releases/2018/12/20
    /establishing-national-bioengineered-food-disclosure-standard.
van der Meer, Piet, Geert Angenon, Hans Bergmans, Hans Jörg Buhk, Sam Callebaut,
    Merijn Camon, Dennis Eriksson, et al. 2021. "The Status under EU Law of Organisms
    Developed through Novel Genomic Techniques." *Eur J Risk Regul* 1 (20). https://doi
    .org/10.1017/err.2020.105.

15

Should We Fear Heritable Genome Editing?

R. Alta Charo

The surprise announcement in November 2018 that a Chinese researcher had implanted and brought to term two gene-edited embryos, resulting in the birth of twin girls, had the effect of galvanizing a debate that goes back decades (Begley 2018; Evans 2002; Kevles 1985). Should we make heritable changes in our children's DNA? Until recently, this question was hypothetical, and the easy response was to say it is too uncertain and unnecessary to be tolerated. Suddenly, however, the possibility that there might be real uses for mitochondrial DNA replacement or for germline editing led to a more nuanced debate, ranging from calls to double-down on prohibiting this technology to discussions of how to permit it for a limited range of conditions, under strict oversight (Baltimore et al. 2015; NASEM 2017; NASEM 2020; Nuffield Council on Bioethics 2018; UNESCO 2015; WHO 2021). Often lacking in this debate has been an effort to look back at debates surrounding earlier advances in reproductive technologies, most of which have been accompanied by fears of eugenics, the loss of human dignity, and the disruption of parent–child relationships. While these advances have each had pockets of abusive uses, they have been integrated into modern life without bringing about wholesale destruction of society.

A true prohibition of germline editing already exists—beginning with the Oviedo Convention, in which 29 countries signed an agreement to prohibit heritable genome editing (HGE). The Oviedo Convention was written specifically to address the intersection of human rights and biomedical developments and aimed to protect the "dignity and identity of all human beings" (Council of Europe 1997). Article 13 reads: "An intervention seeking to modify the human genome may only be undertaken for preventive, diagnostic or therapeutic purposes and *only if its aim is not to introduce any modification in the genome of any descendants*" (emphasis added). In other words, even if done with the best of intentions to ward off devastating—even lethal—conditions, the Convention prohibits any alterations meant to affect descendants, though this position has not been without its critics (Council of Europe 2017; de Wert et al. 2018; Gyngell et al. 2017; Hasson 2018).

Debates around HGE focus on multiple concerns. With respect to physical harm to individuals living in the future, debates involve a risk–benefit analysis complicated by the multigenerational potential of the change (Baylis 2018; Rubeis and Steger 2018). This potential alone introduces questions about the stability and durability of the alteration, its effect under future (presumably different) environments, and the ever-increasing number of generations between the person affected and the person initially giving consent. The International Commission on Human Germline Genome Editing noted in its 2020 report that most diseases are polygenic and that "[s]cientific knowledge is not at a stage at which [heritable editing] for polygenic diseases can be conducted effectively or safely."

Similarly, there is insufficient knowledge to permit consideration of genome editing for other purposes, including nonmedical traits or genetic enhancement, because anticipated benefits in one domain could be offset by unforeseen impact on risk of other diseases." On the other hand, given developments in the underlying science, the same report notes, "[w]ith some notable exceptions, monogenic diseases are individually rare, but together the thousands of monogenic diseases impose significant morbidity and mortality on populations. Current knowledge of medical genetics suggests that the possibility of using [human heritable germline editing] HHGE to increase the ability of prospective parents to have biologically-related children who will not inherit certain monogenic diseases is a realistic one."

A different objection goes directly to how we understand autonomy. As noted in the July 2018 report by the UK Nuffield Council on Bioethics, one might argue that "choosing someone else's genetic endowment . . . offends against the essential dignity and nature of the person as a free and independent human being." This argument is that germline editing interferes with a child's "right to an open future" (Feinberg 1980, 1992). But one response has been that parents not only make many momentous decisions affecting their children's lives, but that the acceptability of parental choices rests on whether they serve to expand or narrow a child's prospects and whether the changes were made for the child's welfare, such as preventing serious disease and disability (NASEM 2017; Nuffield Council on Bioethics 2018).

Of course, it should be noted that many in the deaf community and the community of little people would not define those conditions as disabilities, but rather as varieties of the human community. This view, however, is the exception, and other groups with shared disabilities have not refused the designation, although they often argue that the degree of impairment is as much a function of social and physical context as it is anything intrinsic to the body. Thus, a recurring theme is the need to ensure continued respect for all persons and efforts to make physical structures and social processes accessible to all, while also offering opportunities to use medical interventions to mitigate or eliminate disabling conditions. According to the WHO 2021 report, for heritable editing, these competing goals mean that "[g]ood governance needs to consider both the desires of prospective parents to have genetically connected offspring and the risks to future offspring, as well as possible effects on society, particularly in light of these alternatives."

Indeed, many concerns about germline editing revolve around fear that it will lead to intolerance of imperfection, turn children into commodities rather than the subjects of parental love, and result in stigmatization of those who are disabled (Thiessen 2018). These concerns are familiar. They have been raised repeatedly with each new advance in reproductive technologies, such as prenatal screening, gamete donation, IVF, surrogacy, preimplantation genetic diagnosis (PGD), and cloning. Germline editing is simply the latest rehearsal of what are fundamentally the same concerns around intolerance for diversity or imperfection.

Debates about reproductive technologies and commodification go back at least half a century. By the 1970s, amniocentesis had entered clinical care and was used to screen early second trimester fetuses for chromosomal abnormalities, such as triploidy associated with Down syndrome (Cowan 1993). The prospect of using abortion to eliminate the birth of children with this and other chromosomal conditions sparked widespread discussion about the value persons with disabilities find in their own lives, about the prospect of nongovernmental eugenics (even absent government influence, a pattern of common decision-making among individuals), and about whether the ability to avoid their birth would affect the way parents regard all children and lead to their "commodification," undermining their status as a gift or a blessing. But while the rate of births with Down syndrome has been dramatically affected in some places, the overall number has continued to be substantial, and, if anything, acceptance of people with Down's has only increased in the intervening years (Guralnick, Connor, and Hammond 1995; Hocutt 1996; Kasari et al. 1999; Mansfield, Hopfer, and Marteau 1999; Natoli et al. 2012; Zhang 2020).

The 1980s saw the rising use of artificial insemination, with its associated public fears that women would flock toward the "Genius Bank" for "superior sperm." This did not happen, and reportedly no more than a few dozen children were born from this source. Surveys show that most people simply want donors who will resemble the nongenetic rearing parent (Klock and Maier 1991; Nielsen, Pedersen, and Lauritsen 1995; Nijs 1982; Scheib, Raboy, and Shaver 1998).

That decade was also the era of surrogate motherhood and in vitro fertilization (IVF), two reproductive techniques that were, again, predicted to undermine parent–child relationships. IVF, by which eggs are fertilized in a laboratory and grown until ready for transfer to a woman who will gestate them until birth, was viewed as unnatural, yet parent–child relations have not in fact been harmed. The problematic history of surrogacy is due not to the loss of parental love but rather to the effects of wealth inequality, which has led some to worry about exploitation of low-income women, in the United States and elsewhere (Markens 2007). This risk of exploitation is particularly great when IVF is used in conjunction with surrogacy because, in these cases, the rearing couple uses their own gametes and the pregnant woman has no genetic relationship with the child she bears, so her own physical and genetic characteristics may be of little concern to the rearing couple (*Johnson v. Calvert* 851 P.2d 776 [Cal. 1993]).

The one area where problems have seemed most acute is egg donation. An overly deterministic view of genetics led some rearing couples to seek egg donors with high standardized test scores or academic credentials at elite universities, with exaggerated payments as inducements. While there is little evidence of long-term physical harm from ovulation stimulation, especially when done only once, it still poses some risk to healthy young women. But here, again, despite many articles discussing the phenomenon, evidence of the practice has been anecdotal, and despite many services advertising access to eggs from highly educated women, it is not clear just how widespread demand has become (Almeling 2007; Epstein and Whitehouse 2020).

In the 1990s, the next development to stir controversy was PGD, whereby in vitro embryos could be biopsied and those with known deleterious mutations left unused. Debates surrounding appropriate use of PGD focused on two fears. First, there was concern that parents would use the technique for ever more trivial reasons and finally push society toward the commodification of children and intolerance of imperfection that had been predicted with each of the previous reproductive technology advances (Greely 2016). Here too, however, experience showed that the expense and inconvenience of IVF, plus the limited range of conditions that could be reliably identified, meant that its use was largely restricted to serious or lethal conditions. And the Americans with Disabilities Act (ADA) led to tremendous progress toward making workplaces, homes, and public facilities accessible so that those with disabilities would no longer be isolated from the wider community.

The second fear, however, was that it might be used for sex selection, which in turn was viewed by some as reification of sex differences that had undergirded centuries of discrimination against girls and women. Sex selection had been possible with amniocentesis since the 1970s, but the prospect of selective abortion deterred many people in the United States. PGD, however, offered the opportunity for sex selection without an abortion. Again, only very highly motivated people have been willing to undergo the expense and inconvenience of IVF simply to ensure the birth of a child of one sex or the other (Macklin 2010; Steinbock 2002).

Later in the 1990s, the prospect of human cloning led to a flurry of efforts to ban what was seen as an immoral or dangerous procedure. But the public showed little appetite to pursue it for human purposes, despite some high profile (and rather silly) claims about its success—all of which proved to be fraudulent. Nonetheless, the Oviedo Convention, a host of state laws in the United States, and national policies abroad have adopted criminal penalties for attempting this.

As we entered the twenty-first century, attention turned to mitochondrial replacement, a technique that would be useful for women whose eggs carried mitochondrial DNA known to cause a wide variety of health problems. PGD provided no solution, so for those who wished to maintain a genetic connection to their offspring, one solution was to use donated, healthy mitochondria. Several attempts were made over the course of two decades, beginning in 1996 (Wolf, Mitalipov, and Mitalipov 2015). By 2016, several dozen children around the world had been born after being conceived using this technique, though with varying, and arguably inadequate, attention to long-term follow up and transparency in data reporting and sharing (Institute of Medicine 2016; Kula 2016).

Mitochondrial replacement entered the realm of transgenerational genetic modification, because the altered eggs would result in offspring who, if female, would carry that donated mitochondrial DNA in their own eggs and pass it down to the next generation and so on. The Oviedo Convention appeared to prohibit this process. Critics worried that this technology would become sought-after for older women, including those with no known mitochondrial disease, in the hope it would enhance their fertility. But at least one 2019 study showed no positive effect for fertility in women 37 years of age or older. The experience described suggests that the risks and discomfort (to say nothing of the expense) associated with this technology will likely limit its users to those with a compelling need. In addition, a recent study suggests that the technique is not particularly successful at increasing the chance of conception for older women (Mazur et al. 2019). For the moment, such a procedure has been rendered de facto illegal in the United States by virtue of a federal budgetary provision that precludes FDA review of a request to begin clinical trials involving the technique (Kaiser 2019). The United Kingdom remains the only country with explicit regulations governing permitted indications (Human Fertilisation and Embryology Authority 2015).

In sum, as we approach the end of the first quarter-century of the millennium, a half century of experience with new technologies has been predicted to alter human relations and give people a power they would inevitably abuse, but it has not resulted in these dystopian futures.

Despite this reality, the same predictions are being made about germline editing and have been used to support calls for a moratorium on the practice. In some cases, these calls are tantamount to an indefinite prohibition, because they fail to identify endpoints that would permit the moratorium to be lifted (Spivak, Cohen, and Adashi 2018). In other cases, the conditions outlined as endpoints correlate with many of the conditions outlined by the 2017 NASEM and 2020 International Commission reports, begging the question of how these moratoria would differ from genuine regulation, particularly in countries where public input is made possible by means of both legislative and administrative rulemaking (Lander et al. 2019).

While the debates around germline editing continue and the science remains in preliminary stages, there is still time to take a close look at why and how these earlier reproductive technologies spread, what their limiting factors were, and how such factors might help in formulating better predictions for the range and scale of use one might expect for germline editing. For example, IVF was originally developed to circumvent blocked fallopian tubes, but its use rapidly expanded to encompass idiopathic infertility, male sub-fecundity (in conjunction with intracytoplasmic sperm injection), egg do-nation, and gestational surrogacy. But given the expense, discomfort, and risks of IVF, it did not expand to populations able to conceive through inter-course without significant fear of passing on a serious genetic condition but that (as was feared and predicted at the time) would want to use PGD to screen for conditions that do not seriously impair health. At most, some small uptake involved screening for later-onset cancers. Knowing how much this lack of uptake has been due to a lack of motivating reasons to undergo IVF versus the limits of PGD screening would be quite relevant if one wished to extrapolate to probable patterns of germline use.

It would be worthwhile to calculate the number of people who might be interested in germline editing. Primarily, PGD would not be an option for this small number of people, including couples for whom one parent is homozy-gous for a dominant mutation such as Huntington's. Secondarily, PGD might be an option for couples for whom the number of available embryos follow-ing PGD is quite small. Who are these people, and are their numbers (and their marital patterns) sufficient to raise the specter of exacerbating inequi-ties or creating a genetically superior caste in society, as is feared by some?

A peculiar aspect of the germline editing debate has been the assertion, on one hand, that it could be banned with little cost to reproductive freedom, following the argument that very few people actually need it, and the asser-tion, on the other hand, that it will become sufficiently popular to have a global effect on humanity's own genome.

In the debates surrounding the use of genome editing for germline alter-ation, one of the frequently raised concerns is its effect on human evolu-tion. Given the need to do IVF to provide an embryo on which to perform germline editing, it would seem unlikely to become a sufficiently prominent part of human reproduction to have any evolutionary effects in the near future, even if its substantial technical, regulatory, logistical, and economic barriers could be overcome. Even though many women undergo IVF, it is not a universal practice. Rather, it is used by those with infertility or other con-ditions that make sexual intercourse a risky or inefficient means of reproduc-tion. Even among those using IVF, germline editing would add an element of risk that would not be worth facing in the absence of sufficiently compel-ling motivation. It is for this reason that many fears about the use of germ-line editing for trivial or so-called enhancement purposes are likely to be overblown. More likely to become part of our lives would be genome editing that lacks transgenerational effects, such as somatic editing at the fetal or postnatal stages or—even more limited in its effects—epigenetic editing for transient alterations. For this reason, the World Health Organization appointed an expert committee (WHO 2018) and developed guidance for global governance of genome editing, focusing on a broad range of uses including germline changes (WHO 2021).

If HGE is to proceed anywhere, it must be preceded by rigorous preclinical work, be limited to situations in which it is a reasonable alternative, and be subjected to technical and public oversight. Both the International Commission and the WHO expert advisory committee spoke to these requirements.

From a technical point of view, the International Commission laid out detailed criteria and recommendations. Most importantly, it recommended that no attempt should proceed "unless and until it has been clearly established that it is possible to make precise genomic changes efficiently and reliably without undesired changes in human embryos. These criteria have not yet been met, and further research and review would be necessary to meet them. . . . Clinical use of HHGE should proceed incrementally. At all times, there should be clear thresholds on permitted uses, based on whether a responsible translational pathway can be and has been clearly defined for evaluating the safety and efficacy of the use. . . ."

For any move from research to clinical use, the International Commission also recommended that plans should be in place to evaluate human embryos prior to transfer using:

- developmental milestones until the blastocyst stage is comparable with standard in vitro fertilization practices;
- a biopsy at the blastocyst stage that demonstrates the existence of the intended edit in all biopsied cells;
- no evidence of unintended edits at the target locus; and
- no evidence of additional variants introduced by the editing process at off-target sites.

The International Commission report also included an extensive set of recommendations concerning the conditions that might be serious enough to justify using heritable editing, if other reasonable alternatives did not exist or were, for some reason, unavailable or unacceptable. In its recommendations, it stated that only cases in which the prospective parents' children would inherit the disease-causing genotype for a serious monogenic disease (defined in this report as causing severe morbidity or premature death) could germline editing be clearly justified.

The WHO committee focused on providing advice for developing governance systems for genome editing. Governance was understood to encompass not just formal governmental action, such as legislation and regulation, but also other forms of influence, ranging from conditions on funding to liability or health-care insurance coverage to professional guidelines. The committee also called for the creation of a registry to add transparency to the field by collecting information about ongoing clinical efforts and, especially relevant to heritable editing, about preclinical work being done in laboratories around the world.

Separately, but not entirely unrelated, is the notion of an ecosystem approach to managing rogue actors (Charo 2019). The WHO report outlined a system of collaboration among regulators, professional societies, and research academies around the world to identify and manage research or clinical efforts deemed premature, unsafe, or unethical. The goal would be to forestall exactly the kind of unethical work announced by Dr. He Jiankui at the second international genome editing summit in late 2017. And it would be useful for preventing a repeat of the global experience with fraudulent "clinics" springing up throughout the world offering phony treatments for an assortment of serious ailments advertised as "stem cell therapy" (Charo 2016).

For countries inclined to permit heritable editing, the WHO committee laid out a series of questions the government should be able to answer before concluding that its governance system is adequate. In addition to questions about the technical expertise and capacity available to regulators, the committee listed questions about whether and how to conduct long-term monitoring while maintaining privacy rights, who would determine which genetic concerns would justify such editing, how the country could avoid becoming a magnet for unscrupulous providers, and what medical and financial provisions will be available to anyone harmed by the procedure, now or into the future. Coupled with the technical recommendations of the international commission, the document provides a roadmap for any authority inclined to make such editing legal, while still recognizing that many countries will continue to prohibit it—civilly or criminally—even if safe and effective because of religious or cultural objections.

Good governance for heritable editing should incorporate lessons from past experiences in the use of reproductive technologies, including the scale of patient demand, the equity of access, the safety of the procedures, the effectiveness of governance systems, and the effects on society. As germline editing moves from science fiction to laboratory to (perhaps) the clinic, good governance will always begin with good facts.

REFERENCES

Almeling, Rene. 2007. "Selling Genes, Selling Gender: Egg Agencies, Sperm Banks, and the Medical Market in Genetic Material." *Am Soc Rev* 72 (3): 319–40.

Baltimore, David, Paul Berg, Michael Botchan, Dana Carroll, R. Alta Charo, George Church, Jacob E. Corn, et al. 2015. "A Prudent Path Forward for Genomic Engineering and Germline Gene Modification." *Science* 348 (6230): 36–38.

Baylis, Françoise. 2018. "Counterpoint: The Potential Harms of Human Gene Editing Using CRISPR-Cas9." *Clin Chem* 64: 489–91.

Begley, Sharon. 2018. "Amid Uproar, Chinese Scientist Defends Creating Gene-Edited Babies." *STAT News*, September 28, 2018. https://www.statnews.com/2018/11/28/chinese -scientist-defends-creating-gene-edited-babies/#:~:text=HONG%20KONG%20—%20The %20scientist,International%20Human%20Genome%20Editing%20Summit.

Charo, R. Alta. 2016. "On the Road (to a Cure?)—Stem-Cell Tourism and Lessons for Gene Editing." *N Engl J Med* 374 (10): 901–03.

Charo, R. Alta. 2019. "Rogues and Regulation of Germline Editing." *N Engl J Med* 380: 976–80.

Council of Europe. 1997. "Oviedo Convention and Its Protocols." https://www.coe.int/en /web/bioethics/oviedo-convention.

Council of Europe. 2017. "Recommendation 2115: The Use of New Genetic Technologies in Human Beings." https://assembly.coe.int/nw/xml/XRef/Xref-XML2HTML-en .asp?fileid=24228&lang=en.

Cowan, Ruth Schwartz. 1993. "Aspects of the History of Prenatal Diagnosis." *Fetal Diagn Ther* 8 (suppl. 1): 10–17.

de Wert, Guido, Guido Pennings, Angus Clarke, Ursula Eichenlaub-Ritter, Carla G. van El, Francesca Forzano, Mariëtte Goddijn, et al. 2018. "Human Germline Gene Editing: Recommendations of ESHG and ESHRE." *Hum Reprod Open* 1 (1): 445–49.

Epstein, Sonia F., and Polina N. Whitehouse. 2020. "Inheriting the Ivy League: The Market for Educated Egg and Sperm Donors." *Crimson*, April 30, 2020. https://www.thecrimson .com/article/2020/4/30/inheriting-the-ivy-league/.

Evans, John H. 2002. *Playing God? Human Genetic Engineering and the Rationalization of Public Bioethical Debate*. Chicago: University of Chicago Press.

Feinberg, Joel. 1980. "The Child's Right to an Open Future." In *Whose Child? Children's Rights, Parental Authority, and State Power*, edited by William Aiken and Hugh LaFollette, 124–53. Totowa, NJ: Rowman & Littlefield.

Feinberg, Joel. 1992. *Freedom and Fulfillment*. Princeton: Princeton University Press.

Greely, Henry T. 2016. *The End of Sex and the Future of Human Reproduction*. Cambridge: Harvard University Press.

Guralnick, Michael J., Robert T. Connor, and Mary Hammond. 1995. "Parent Perspectives of Peer Relationships and Friendships in Integrated and Specialized Programs." *Am J Ment Retard* 99 (5): 457–76.

Gyngell, Christopher, Marie Sklodowska-Curie Fellow, Thomas Douglas, and Julian Savulescu. 2017. "The Ethics of Germline Gene Editing." *J Appl Philos* 34 (4): 498–513.

Hasson, Katie. 2018. "UK's Nuffield Council Releases Report on Human Genome Editing." *Bio Political Times*, August 2, 2018.

Hocutt, Anne M. 1996. "Effectiveness of Special Education: Is Placement the Critical Factor?" *Future Child* 6 (1): 77–102.

Human Fertilisation and Embryology Authority. 2015. *Human Fertilisation and Embryology (Mitochondrial Donation) Regulations*. Adopted October 29, 2015. http://www.legislation.gov.uk/uksi/2015/572/contents/made.

Institute of Medicine. 2016. *Mitochondrial Replacement Techniques: Ethical, Social, and Policy Considerations*. Washington, DC: National Academies Press.

Kaiser, Jocelyn. 2019. "Update: House Spending Panel Restores U.S. Ban on Gene-Edited Babies." *Science*, June 5, 2019. https://doi.org/10.1126/science.aay1607.

Kasari, Connie, Stephanny F. N. Freeman, Nirit Bauminger, and Marvin C. Alkin. 1999. "Parental Perspectives on Inclusion: Effects of Autism and Down Syndrome." *J Autism Dev Disord* 29: 297–305.

Kevles, Daniel J. 1985. *In the Name of Eugenics: Genetics and the Uses of Human Heredity*. Berkeley: University of California Press.

Klock, Susan C., and Donald Maier. 1991. "Psychological Factors Related to Donor Insemination." *Fertil Steril* 56 (3): 489–95.

Kula, Shane. 2016. "Three-Parent Children Are Already Here." *Slate*, February 18, 2016. https://slate.com/technology/2016/02/three-parent-babies-have-been-here-since-the-late-90s.html.

Lander, Eric, S., François Baylis, Feng Zhang, Emmanuelle Charpentier, Paul Berg, Catherine Bourgain, Bärbel Friedrich, et al. 2019. "Adopt a Moratorium on Heritable Genome Editing." *Nature* 567, 165–168.

Macklin, Ruth. 2010. "The Ethics of Sex Selection and Family Balancing." *Semin Reprod Med* 28 (4): 315–21.

Mansfield, Caroline, Suellen Hopfer, and Theresa M. Marteau. 1999. "Termination Rates after Prenatal Diagnosis of Down Syndrome, Spina Bifida, Anencephaly, and Turner and Klinefelter Syndromes: A Systematic Literature Review." *Prenat Diagn* 19 (9): 808–12.

Markens, Susan. 2007. *Surrogate Motherhood and the Politics of Reproduction*. Berkeley: University of California Press.

Mazur, Pavlo, Lada Dyachenko, Viktor Veselovskyy, Yuliya Masliy, Maksym Borysov, Dmytro Mykytenko, Valery Zukin, et al. 2019. "P-221 Mitochondrial Replacement Therapy Give No Benefits to Patients of Advanced Maternal Age." Paper presented at American Society for Reproductive Medicine (ASRM) Scientific Conference and Expo, Philadelphia, October 14–16, 2019. https://asrm.confex.com/asrm/2019/meetingapp.cgi/Paper/2347.

National Academies of Sciences, Engineering, and Medicine (NASEM). 2017. *Human Genome Editing: Science, Ethics, and Governance*. Washington, DC: National Academies Press.

National Academies of Sciences, Engineering, and Medicine (NASEM). 2020. *International Commission on the Clinical Use of Human Germline Genome Editing*. Washington, DC: National Academies Press.

Natoli, Jaime, Deborah L. Ackerman, Suzanne McDermott, and Janice G. Edwards. 2012. "Prenatal Diagnosis of Down Syndrome: A Systematic Review of Termination Rates (1995–2011)." *Prenat Diagn* 32 (2): 142–53.

Nielsen, Anders Faurskov, Bjørn Pedersen, and Jørgen Glenn Lauritsen. 1995. "Psychosocial Aspects of Donor Insemination: Attitudes and Opinions of Danish and Swedish Donor Insemination Patients to Psychosocial Information Being Supplied to Offspring and Relatives." *Acta Obstet Gynecol Scand* 74 (1): 45–50.

Nijs, Piet. 1982. "Aspects médico-psychologiques de l'insémination artificielle." *Jus Medicum* (Acta Fourth World Congress on Medical Law) 69–81.

Nuffield Council on Bioethics. 2018. *Genome Editing and Human Reproduction.* London: Nuffield Council. https://www.nuffieldbioethics.org/publications/genome-editing -and-human-reproduction.

Parliamentary Assembly. 1997. *Convention for the Protection of Human Rights and Dignity of the Human Being with Regard to the Application of Biology and Medicine.* Strasbourg: Council of Europe. https://www.coe.int/en/web/bioethics/oviedo-convention.

Rubeis, Giovanni, and Florian Steger. 2018. "Risks and Benefits of Human Germline Genome Editing: An Ethical Analysis." *Asian Bioeth Rev* 10 (2): 133.

Scheib, Joanna E., B. L. Raboy, and Phillip R. Shaver. 1998. "Selection of Sperm Donors: Recipients' Criteria and Donor Attributes that Predict Choice." *Fertil Steril* 70 (suppl. 1): S279.

Spivak, Russell, I. Glenn Cohen, and Eli Y. Adashi. 2018. "Moratoria and Innovation in the Reproductive Sciences: Of Pretext, Permanence, Transparency, and Time Limits." *J Health Biomed Law* XIV: 5–26.

Steinbock, Bonnie. 2002. "Sex Selection: Not Obviously Wrong." *Hastings Cent Rep* 32 (1): 23–28.

Thiessen, Marc A. 2018. "Gene Editing Is Here: It's an Enormous Threat." *Washington Post*, November 29, 2018. https://www.washingtonpost.com/opinions/gene-editing-is-here -its-an-enormous-threat/2018/11/29/78190c96-f401-11e8-bc79-68604ed88993_story .html.

United Nations Educational, Scientific and Cultural Organization (UNESCO). 2015. "UNESCO Panel of Experts Calls for Ban on 'Editing' of Human DNA to Avoid Unethical Tampering with Hereditary Traits." https://en.unesco.org/news/unesco -panel-experts-calls-ban-editing-human-dna-avoid-un- ethical-tampering-hereditary -traits.

Wolf, Don P., Nargiz Mitalipov, and Shoukhrat Mitalipov. 2015. "Mitochondrial Replacement Therapy in Reproductive Medicine." *Trends Molec Med* 21 (2): 68–76.

World Health Organization (WHO). 2018. "Announcement of Expert Panel." https://www .who.int/ethics/topics/human-genome-editing/en/.

World Health Organization (WHO). 2021. "Expert Advisory Committee on Developing Global Standards for Governance and Oversight of Human Genome Editing." https:// iris.who.int/bitstream/handle/10665/342484/9789240030060-eng.pdf?sequence=1.

Zhang, Sarah. 2020. "The Last Children of Down Syndrome." *Atlantic*, December 12, 2020. https://www.theatlantic.com/magazine/archive/2020/12/the-last-children-of -down-syndrome/616928/.

16

Advancing Progressively Backwards

Guiding and Governing Heritable Genome Editing

J. Benjamin Hurlbut

What have we to do
But stand with empty hands and palms turned upwards
In an age which advances progressively backwards?

—T. S. Eliot

Note: Throughout this article, I have redacted many of the identities of Dr. He Jiankui's interlocutors. My aim is not to impugn these individuals' reputations but to give a picture of the scientific milieu they are part of and have helped to engender. They know who they are. It is their responsibility to step forward, explain themselves, and shoulder responsibility for reform. I invite them to do so, and I hope that readers of this essay will join me in that invitation.

In November 2018, a Chinese scientist named He Jiankui caused an international uproar by announcing the birth of two babies whose DNA he had edited. The news, which broke on the eve of the Second International Summit on Genome Editing in Hong Kong, elicited swift condemnation from an apparently univocal international scientific community—a community that then proceeded to claim sovereign authority over how the future of heritable genome editing (HGE) should be governed.

Three international summits have been held and a bookshelf of advisory reports have been produced. Yet the fundamental challenges of governance have yet to be seriously confronted through the kind of thorough, open, and inclusive deliberation they demand.

The ability to make direct genetic changes to the DNA of future children poses profound challenges for governance: Should it be done? For what purposes and subject to what limitations? No less importantly, who should decide? Genome editing is one of a growing body of technologies with the potential to alter what it means to be human. How should we as a human community guide and govern this emerging technology?

The question of governance is not only about whether HGE should be done, but about how societies should approach that question—through what processes; in what terms; drawing upon what forms of expertise and moral imagination; and guided by what priorities, interests, and ethical commitments (Hurlbut 2019). Modes of deliberation and governance shape what questions are asked and what questions go unasked, what perspectives are privileged or silenced in public debate, and what priorities and practices come to be embedded in our institutions (Hurlbut 2017a). Present approaches to governance will not only shape the trajectory of genome editing, but also the ways social and political communities will contend with other technologies that touch upon fundamental dimensions of human life in the future.

The root of the word *governance* derives from an ancient Greek nautical term. It meant "to steer the ship." That is a meaning worth recovering. In reflecting upon modes of governance, we must look beyond systems of codified rules and mechanisms of oversight to ask more fundamental questions: what course are we traveling, charted by whom, navigating with what instruments, who is at the helm, and what winds and currents are bearing us along—or driving us off course? Understood in this way, governance extends to the ways problems and issues are framed, terms of debate are set, and particular ways of thinking and speaking come to be privileged over others. These dimensions of framing and language also affect notions of who has the expertise, authority, and responsibility to govern.

This concluding chapter examines some dominant ways of thinking and speaking to illuminate how these largely unexamined dimensions of discourse steer the ship of science and technology governance in consequential ways. I examine the causes and consequences of He Jiankui's ill-begotten experiment, drawing on extensive interviews I conducted with him and his associates as the controversy surrounding the "CRISPR babies experiment" unfolded. I show how He's experiment gave expression to an ethos in science that celebrates risk-taking, "disruptive" innovation, and the imperative to race ahead in the name of progress. This ethos not only shapes the culture of technoscience itself, but also the contours of governance. It underwrites a vision of science as free, autonomous, and self-governing—what I call "sovereign science" (Hurlbut 2017a,b)—and of society and its institutions of governance as always lagging behind and thus incapable of guiding or governing the technological future. When He undertook his transgressive experiment, he drew his authorization and moral warrant from this vision of scientific sovereignty.

Rightly understood, the He affair is not so easily dismissed as an aberration from scientific norms. It poses hard lessons that must be faced as we contend with a technology with the power to alter what it means to be human.

The Unthinkable Has Become Conceivable

In December 2015, the US National Academy of Sciences and the National Academy of Medicine, with the Royal Society and the Chinese Academy of Sciences, convened the first International Summit on Human Gene Editing in Washington, DC. Nobel laureate and former Caltech president David Baltimore opened the Summit by explaining why it is urgent to address the implications of human genome editing. Although genetic engineering techniques had been around for decades, they were imperfect, and applying them to the human germline was "initially unthinkable." But with advances in genome editing, "the unthinkable has become conceivable. . . . Now we must face the questions that arise: how, if at all, do we as a society want to use this capability?" (Baltimore 2015).

Baltimore's statement reprises a familiar sequence: scientific knowledge generates technological applications which, in turn, produce societal impacts and consequences. From science to technology to society: this is the "linear model" of innovation. Scholars of innovation generally reject the linear model. Innovation tends to follow a far more complex and circuitous pathway, shaped by its surrounding social, legal, institutional and economic environment (Edgerton 2017; Narayanamurti and Odumosu 2016).

Yet the linear model looms large in discourse about the governance of technology, particularly in ethically fraught domains like genome editing. The *idea* that technological innovation follows a linear path shapes bioethical deliberation and biotechnology governance by treating "basic" science and its potentially ethically problematic societal applications as distinct. In this sense the linear model is more a model of governance than of innovation. Framing technological change in linear terms has several important consequences.

First, the widespread assumption, replicated in Baltimore's statement above, holds that ethical issues arise as a downstream consequence of development of particular technologies: advances in science and technology raise "ethical questions." Science comes first, then technology, then ethical considerations. The effect of such thinking is to treat ethical considerations as subsidiary and temporally subsequent to technological change. Ethical concerns are meaningful only once a technology exists or will in the near future. This science-first approach positions scientific experts as best qualified to play gatekeepers to ethical deliberation. They define what technological potentials are realistic and, thus, which scientific and technological developments do or do not warrant ethical evaluation. In short, they decide when "the unthinkable has become conceivable," such that it is reasonable to react to its ethical implications. This way of thinking positions ethics not as guiding and governing scientific and technological change but as reactive to it. Ethical consideration comes—and, in this way of thinking, should only come—at the threshold of transition from technology to society.

Positioning ethics after innovation delimits a zone of science that ought not be subjected to ethical scrutiny or interference because it is ostensibly not yet itself impacting society. Thus, so-called basic research tends to be treated as lying outside the scope of ethical evaluation, and thus as the sovereign territory of scientific self-rule. This is so even when the primary purpose of research is to develop a technology whose use is ethically questionable, for instance genetic editing of human embryos where the intention is to study the process but not (yet) use the embryos to initiate a pregnancy. Indeed, most discussions of HGE have focused on the movement of that technology into society—"clinical applications"—leaving aside the question of whether research that drives in this direction is itself ethically problematic. In the hundreds of pages of reports generated since 2015, there are only a few minor exceptions to this rule (e.g., European Group 2016; Lanphier et al. 2015).

This linear model of governance profoundly constrains ethical deliberation, even as it affords excessive authority to scientists to set research and innovation agendas. Yet the question of whether it is right and appropriate to modify the genomes of future children transcends the state of technology at any given moment. It goes to the question of what dimensions of human life ought not be taken as objects of manipulation and control, regardless of what technological means do or do not presently exist. Those questions were already a focus more than half a century ago when debates about human genetic engineering focused broadly on the relationship between emerging scientific horizons and the identity, integrity, and perfectibility of the human (Evans 2002). In 1969, Robert Sinsheimer, whose research was foundational for the development of biotechnology, celebrated "defined genetic improvement of man" as a means "to carry on and consciously perfect" the human species (141). In 1962, Julian Huxley envisioned powers for directed human evolution that raise—and must be guided by—the question: "What are people for?" (Huxley 1963). And in 1980, following the advent of recombinant DNA and in vitro fertilization, three major religious organizations called upon President Jimmy Carter's administration to attend to the "new era of fundamental danger triggered by the rapid growth of genetic engineering" and to associated questions of "the fundamental nature of human life and the dignity and worth of the individual human being" (President's Commission 1982).

Yet beginning in the 1980s with the advent of professional bioethics, there was a shift toward a more narrow approach to bioethics that increasingly followed the reactive model described above. The effect was to delimit ethical deliberation about human genetic engineering to the immediate technological environment, focusing only on the next step, not on where the path ultimately leads. The effect of this incrementalist approach has been to delay ethical inquiry until the technology is already upon us—until the unthinkable has become conceivable.

This move has also had the effect of devaluing broader, autonomous inquiry into the value of the human, independent of scientific and technological developments. It marginalizes deliberation about what concepts of human integrity, well-being, and progress should guide biotechnological development. Instead, it has made technologies like CRISPR the primary focus of an always merely reactive ethical debate. It fetishizes technology, making technological novelties the centerpiece of ethical deliberation, where that deliberation is itself conditioned by corollary expert judgments about which technological scenarios are, in Baltimore's words, "conceivable" and thus warrant ethical consideration (Hurlbut 2018).

A second effect of this science first, ethics second approach is that bright lines drawn *before* scientists can cross them tend to be treated as obsolete and irrelevant once they can be crossed. For instance, the National Academies of Sciences, Engineering, and Medicine (NASEM) report on *Human Genome Editing: Science, Ethics, and Governance* (2017a) broke with a longstanding prohibition against intentional modification of the human germline, arguing that for certain applications, germline was acceptable if proven safe. Although this was a significant ethical reversal, committee cochair Richard Hynes explained it as a straightforward result of changes in the state of science and technology: it meant nothing to prohibit HGE when it was not yet possible to do it, but now that it is, there is no need to take the earlier prohibition seriously.

At the event celebrating the release of the report, Hynes noted that "There has been a line drawn by many that says . . . you should refrain [from modifying the human germline]. That was mostly because there was no way of considering how to do that at all . . . so nobody was arguing that it should be done." In this approach, the capacity for technological intervention defines the horizon of human ethical reflection: if it cannot be done, it is meaningless to say that it should not be done. Only once it can be done is it no longer premature to ask whether it should be done. And yet, the same actors who declare that questions are no longer premature often at the same time assert that it is too late to ask them.

That is because the notion of prematurity tends to go hand in hand with a corollary discourse of inevitability. This is the third consequence of the science first, ethics second paradigm. Not only is the preexisting prohibition treated as meaningless because it was "premature," but prohibition is itself often marked as impossible: once the technology exists, its use is inevitable. It is too late to stop it. The genie is out of the bottle.

From 2015 forward, the notion that HGE is inevitable figured prominently in public discourse (along with the corollaries that we had better get used to it, accept it, and regulate it, but not question whether we should have it). For instance, a May 2015 interview of scientific luminaries conducted by *Nature Biotechnology* began with the question, "With the current pace of advances in the use of gene editing technology, IVF and germ stem cell research, to what extent do you think germline engineering is inevitable?" (Bosley et al. 2015). Numerous prominent figures in genome editing science and bioethics—Jennifer Doudna, Robin Lovell-Badge, Dana Carroll, Hank Greely, Craig Venter, and Tony Perry, among others—answered that it was. Some of these figures have since played (and continue to play) major roles in shaping deliberation and governance of HGE.

The widely shared presumption that HGE is inevitable shifts focus from asking whether we want it—and, no less importantly, how we as a human community should seek to answer this question—to treating it as a fait accompli that we will have it (Jasanoff and Hurlbut 2018). Asserting inevitability is neither innocent nor inconsequential. When news of the world's first "CRISPR babies" broke, the predictions that it was inevitable seemed prophetic. Yet the declaration of inevitability, made by prominent scientists empowered to shape the future, was, in fact, a self-fulfilling prophecy.

Breaking the Glass

On November 25, 2018, the eve of the Second International Summit on Human Genome Editing in Hong Kong, the world learned that a scientist at the Southern University of Science and Technology in Shenzhen, China, had edited human embryos and transferred them into their mother's womb, resulting in the birth of twin girls (Regalado 2018). It was later learned that another mother was pregnant with (and ultimately gave birth to) a third genetically edited child. The target of the edits was *CCR5*, a gene associated with susceptibility to HIV. The scientist, He Jiankui, had attempted to induce a deletion in that gene that, he hoped, would render the babies immune to HIV. He's experiment received swift and fierce condemnation from the international scientific community. He was immediately labeled a "rogue" who violated not only his research subjects but the integrity of science itself.

The rogue narrative offers a neat and clean story. It draws a bright line between the *responsible* and the *rogue*, the *community* and the *outlaw*, and the *us* and the *them*. It places blame outside the scientific community, externalizing failure by attributing it to one individual's aberrant behavior and thereby limiting damage to the community's credibility. But that story is false. He's ill-conceived project was not a simple aberration from the norms of international science. Rather, it was shaped and influenced by them, and in certain respects was a direct reflection of them. Whereas the rogue narrative invites righteous condemnation, the more complex story should foster humility in self-reflection.

The rogue narrative took shape in response to the unexpected, public revelation of He's experiment. It has been told definitively, unequivocally, and consequentially by a range of narrators. Even some extended and influential examinations have taken that narrative at face value, neither questioning nor investigating its underlying assumptions (Greely 2021; Isaacson 2021; see also Hurlbut 2021). By contrast, the account that follows is based on extensive conversations I had with He and others from his lab and community of colleagues starting in December 2018. It offers a picture of He's motivations that implicate a wider scientific community than the rogue narratives allow for, and which thereby challenges overly simplistic approaches to governance that focus on stopping rogues while setting the putatively responsible free. The purpose of my analysis is not to excuse or defend He, because I consider what he did to be profoundly unethical (Hurlbut, Jasanoff, and Saha 2018). But in the rush to condemn, few in the scientific and bioethics communities have bothered to inquire why He did what he did and how his actions were connected to a larger scientific context and community, let alone whether his actions hold broader lessons.

He Jiankui was driven by the high-octane milieu of contemporary biotechnology, both in the United States and in China. He internalized ideas that led him to believe that his experiment would elevate his status in the international scientific community, advance his country in the race for scientific and technological dominance, and drive scientific progress forward against the headwinds of ethical conservatism and public fear. Thus, far from rejecting the norms of his professional community, the ideas that influenced He are at once powerful and ubiquitous in the world of science.

Indeed, He's motivations are so familiar that they are almost mundane. He, like many scientists, aspired to international recognition for his research, and he therefore chose a research area that was at once cutting edge and controversial. Like many scientists, He aspired to do pioneering work that would gain him ownership of a new technology area—reputationally and commercially. And, like many scientists, he had a patriotic desire to advance the standing of his country in international science.

I interviewed He when he was under house arrest in December 2018. He recounted the thinking that led him to act. He felt Chinese science was regarded as second-class relative to American science and recounted the disgust he felt when prominent American scientists had roundly criticized Huang Junjiu's 2015 experiment in which he edited nonviable human embryos that were donated for research by a patient undergoing fertility treatment (Liang et al. 2015). By contrast, when Shoukhrat Mitalipov of Oregon Health and Science University conducted a germline genome editing experiment less than two years later, he created many viable human embryos for research specifically in order to edit them (Ma et al. 2017). In contrast to Huang, Mitalipov's much more controversial experiment earned him recognition and valorization—in the media, at his university, in *Nature*, and more. (*Nature*, which published Mitolipov's paper, had rejected Huang's paper, partly over "ethical concerns;" see Cressey and Cyranoski 2015.)

He Jiankui saw this as a double standard: he told me it was "scientific racism against the Chinese." He wanted to challenge this asymmetrical treatment of Chinese science by demonstrating that a Chinese researcher could do pathbreaking work ahead of the Western competition. In his application to the Chinese ethics committee that reviewed and approved his experiment, he wrote: "In this ever more competitive global pursuit of applications for gene editing, we hope to be a stand-out" (He 2018). Behind He's aspiration was the corollary notion—espoused by the leading voices in the field to whom He listened attentively—that HGE was inevitable. It was going to happen. Some scientist in some country would play pioneer. So, he thought, why not him?

Thus, the narrative of inevitability was a self-fulfilling prophecy. He acted on the presumption that the action would be taken sooner or later by someone. Indeed, he felt an urgency to act lest he lose the all-significant scientific position of being the first in a field for which there are no prizes for second place, where first is valorized and rewarded by the most coveted of accolades, from Nobel Prizes to patents.

Inevitability also functioned as a kind of authorization, a displacement of responsibility from He's shoulders as the risk-taking pioneer. If the future was a foregone conclusion, a view widely shared and often repeated by his prominent senior colleagues, all He felt he had control over was whether that future would be claimed by him on behalf of his country, or by someone else on behalf of theirs.

I met He at a small meeting at the Innovative Genomics Institute (IGI) at the University of California–Berkeley in January 2017. At that meeting, He encountered a variety of prominent scientists in genome editing and adjacent fields. It would prove to be highly influential for him, though in ways that its organizers did not intend and may remain unaware of. At the meeting, He—who was by far the most junior scientist in the room— mostly sat quietly and listened, gathering ideas and taking cues from his eminent senior colleagues. Reflecting on that meeting almost two years later, he recalled a particular statement that influenced him deeply: "[A very senior scientist] said 'many major breakthroughs are driven by one or a couple of scientists . . . cowboy science.' That strongly influenced me . . . you need a person to break the glass." It was a statement that also struck me: I recorded it verbatim in my notes as an instance of that hubristic imagination wherein power to produce progress, and thus the responsibility to guide and govern it, are the monology of science. He, by contrast, took it as an authorization to appoint himself the courageous scientific agent who would assume the risk necessary to breakthrough to an inevitable future.

Following the meeting, He began reading about great scientific figures— Thomas Edison, Edward Jenner, Christiaan Barnard and, most importantly, Robert Edwards. Forty-five years ago, Edwards undertook a risky experiment without public knowledge or oversight that produced a baby through unprecedented technological means. He revealed his achievement—the first "test-tube baby"—not through a scientific journal but in the popular press. It was enormously controversial at the time. Thirty years later, Edwards was awarded the Nobel Prize in medicine. From this and other heroic stories of scientific risk-taking—Barnard's 1967 heart transplant, Jenner's vaccination of eight-year-old James Phipps—He became convinced of what he had heard at the IGI meeting: major breakthroughs are driven by one or a couple of risk-taking scientists. Heroic scientific achievements are often initially controversial; someone must break the glass.

He also investigated how long it takes for controversial science to be recognized and valorized as pioneering science. He gathered his observations in a PowerPoint he presented to his lab: "Christiaan Barnard: Domestic, 3 weeks, International, 1 Year; Robert Edwards: 7 years; Edward Jenner: 1 year." He imagined himself in the image of these heroic scientific pioneers: initially marked by controversy, but ultimately celebrated. From them he drew the courage to take a controversial first step—breaking the glass—in the name of pushing science, and humanity, forward.

I want to draw attention not only to He's misguided dream of personal celebrity, but to the celebratory narrative he drew upon in developing that dream. He found justification and authorization in the story about the relationship between scientific and social change that sits at the core of the contemporary scientific imagination: science races ahead, and society (and law, ethics, and regulation) inevitably lags behind.

This story is often told and rarely questioned. At the release event for the National Academies consensus report on *Human Genome Editing: Science, Ethics and Governance* in 2017, National Academy of Sciences (NAS) President Marcia McNutt said: "As is always the case, the speed at which the science is advancing outpaces society's ability to grasp its implications" (NASEM 2017b). Her comment is just one example of a virtually endless supply in contemporary discourse around science and technology. The lag narrative is part of the cultural vernacular. Technology is moving so fast that society cannot keep up. "Rapidly moving science," "disruptive technology"—these ideas position science and technology (and those who drive them) as the future-makers and society and its institutions as passively carried along. This construction recapitulates the narrative of linear development discussed above—from science to technology to society—and positions ethics and democratic governance as inevitably laggard and, therefore, merely reactive.

The lag narrative is false. Science and technology do not spring forth mature into naïve social space like Athena from the forehead of Zeus. Law, norms, and cultural environments profoundly shape the forms science and technology take. But the lag narrative has social effects nevertheless because it is believed and deferred to. It is a self-fulfilling prophecy: in asserting that society is impotent to govern science because science is always ahead and society is always behind, it authorizes science to govern itself. If science necessarily outpaces law and ethics, legal and ethical reform *must* defer to science and follow its lead. This narrative also positions science as the arbiter of progress. The corollary to self-governing science is the specter of "overregulation": the idea that if society attempts to overcome the lag by preemptively imposing constraints on innovation, it will inhibit innovation and deny society its benefits. So, society lags behind and must lag behind—that is, society must restrain itself from trying to get ahead because to do so will inhibit science's ability to anticipate and realize the social benefits of innovation.

This notion of vanguard science and laggard society underwrites the norm of scientific autonomy—the sovereignty of the "republic of science" (Hurlbut 2017b; Polanyi 1962). Ethics and law must not only play catch-up, but they also must not try to get ahead lest they interfere with science and inhibit innovation. Thus, what is characterized as a natural and necessary temporal disjunction between science and ethics—innovation first, ethics second—is, in practice, a way of claiming the right to self-governance, the sovereign authority to captain the ship. Because science inevitably leads and ethics inevitably lags, science *must* lead, and ethics must follow. Thus, the lag narrative encodes a powerful imperative of governance, grounded in a false account of sociotechnical change: to achieve progress for society, science must be self-governing, including in rendering judgment about what should be done in the name of progress. Thus, it positions sovereign science not only as the driver of technological progress but also of cultural and moral progress. In short, the lag narrative imbues the scientific and technological vanguard with elevated moral authority, positioning risk-taking science as the standard-bearer of progress.

Like much of his professional community, He internalized the lag narrative and bought into the idea that scientists have an affirmative responsibility to push forward into controversial domains to advance science and "accelerate" corollary social and ethical realignments. He told me: "If we are waiting for society to reach a consensus . . . it's never going to happen. But once one or a couple of scientists make first kid, its [sic] safe, healthy, then the entire society including science, ethics, law, will be accelerated. Speed up and make new rules So, I break [sic] the glass."

He reached for this convenient narrative to justify acting. He thereby appropriated the authority to govern the future. Yet in this regard, He is far from unique. This story of science-driven progress is one that scientists and technologists often tell to arrogate to themselves the authority to render judgment about what technological futures are desirable (Jasanoff 2019). The story authorizes them to supersede, ignore, or even to violate society's norms when they run counter to science's narrative of progress.

This is an illegitimate exercise of power masquerading as an imperative of governance. It is an expression of hubris hidden behind a story of inevitable change: science races ahead, society lags; science drives progress; new technologies render old rules, norms, and ethical commitments obsolete; and society cannot halt or reverse technological progress—all it can do is adapt.

But let us be clear: these dynamics are not natural or inevitable. They are asserted and enacted. They are not intrinsic features of technological change. They are instruments of appropriated governance.

We can look once again to the He affair to see the dynamic of appropriation in action. Much of the media coverage of He's ill-conceived experiment has focused on who knew about it in advance and whether they erred in failing to blow the whistle (Belluck 2019). Initiatives for reform after the He debacle include mechanisms for members of the scientific community to report wrongdoing (WHO 2019). But such mechanisms will be effective only if scientists' actions are seen as wrongdoing. It is a mistake to presume that all who knew about He's project opposed it. The individuals who were strongly critical represent only a small minority of the people He consulted. Judging from my conversations with He and others, which are corroborated by a significant body of documentary evidence, He consulted with dozens of people, including well-known and respected American scientists. Most expressed support. In most cases, these supporters knew about the actual project. A small number affirmed HGE and the idea that some maverick would have to be the first to cross the Rubicon, but expressed reservations about the genetic target He had selected. Only four expressed unequivocal opposition. Indeed, He told me that one of his interlocutors, an accomplished genome-editing scientist at a leading US university, told him that when He commercialized his baby-making enterprise (because that is what one is expected to do with one's inventions in the world of biotechnology), he wanted to be the company's fifth employee. He did not want to be the second, or even the third or fourth, because the technology would still be controversial at that stage. He wanted to wait for the company to be developed enough to be established and accepted. That is, he wanted to wait for society to catch up.

In sum, far from "going rogue" and rejecting the norms and expectations of his professional community, He Jiankui was guided by them. He believed his work would impress his colleagues because some of them told him as much. Thus, the "rogue" label rightly belongs not to He, the lone scientist, but to a scientific community that asserts that it alone has the authority to determine what it should and should not do. Unmoored from society's priorities and norms, self-governing science is rogue science. It is rogue not because it has abandoned its commitment to serving the good of society. It is rogue because it has appropriated from society the authority to declare what is good, dismissing ethical dissent as an ephemeral effect of a lagging society.

Affairs of State

Just as He had the encouragement and support of numerous Western scientific colleagues, he was also encouraged by members of the Chinese scientific community and government. They likewise bought into the idea of an international race to scientific and technological primacy.

In a 2016 speech on science and technology, which was essentially a declaration of national innovation policy, Xi Jinping declared:

> To take hold of science and technology innovation, we cannot wait and see; we cannot blindly follow where others are leading. We must seize every minute. . . . Do not let scientists be bound by red tape. . . . Do not let the efforts of scientists be delayed by endless reports and approvals. . . . We must unite together . . . and advance our goal of building China into the world's leader in science and technology.

Shortly before Xi spoke those words, He had been recruited to a faculty position at the Southern University of Science and Technology (SUSTech) in Shenzhen, China. SUSTech presents itself as "the Stanford of China"—nimble, entrepreneurial and at the center of China's premier innovation enclave. He had been recruited back to China through the Thousand Talents program and had become the face of SUSTech entrepreneurialism. One of his companies, Direct Genomics, was seen as a model of Chinese biotechnology innovation. He and Direct Genomics had been featured in the Chinese national media. Former Secretary of State John Kerry had visited He at Direct Genomics offices. Nobel Laureate Craig Mello was on Direct Genomics board (and served as an erstwhile advisor to He on his CRISPR babies project, in spite of his modest misgivings about it). He and his company had been featured in *Nature Biotechnology*. And when the CRISPR babies story broke in November 2018, the city of Shenzhen was in the process of installing He's picture and profile on a column in a public park. It was to be part of a public art installation made up of a cluster of columns, each featuring a different Shenzhen technology innovator-celebrity—the pillars of the Chinese high-tech community.

In undertaking the HGE project, He was looking to take his contributions to Chinese innovation to the next level. He explained to me that he was uninspired by the prospect of just starting more companies and making more money. He wanted to do something more consequential for both science and country. As he was beginning to make a place for his HGE project, he sought the advice and blessing of several influential Chinese officials. These included, among others, Zhu Qingshi, the former president of SUSTech, who had also been the president of the University of Science and Technology of China, He's undergraduate institution; and Liu Huafung, a high-ranking official in the Organization Department of the Chinese Communist Party.

The Organization Department is a core apparatus of the Chinese Communist Party; it controls personnel appointments to the party, and thus wields enormous power. As Liu was responsible for cultivating talent in science and technology, he was a particularly powerful ally for He. Both Zhu and Liu had full knowledge of He's project from the very beginning, and both were strongly supportive. From these and other well-connected supporters, and the enthusiastic affirmations he received from the range of Chinese scientists and officials he consulted, He drew reassurance that his work, even if controversial, would be well-received.

As the project unfolded, He found additional support. In the summer of 2018, he paid a visit to John Zhang, a Cambridge-trained physician-embryologist who runs a large and highly successful fertility clinic in New York City called New Hope Fertility. Zhang produced the first "three parent baby" by taking manipulated embryos to his outpost clinic in Guadalajara, Mexico—beyond the reach of US law that prevented him from inducing a pregnancy with the manipulated embryos on American soil. He and Zhang quickly formed a partnership and began exploring opportunities in China for establishing a commercial operation for HGE and three-parent embryo services. In October 2018, they met with the vice governor of the Chinese province of Hainan. Hainan is home to the Boao medical tourism campus, a special regulatory zone that is liberated from the "red tape" and "reports and approvals" of the Chinese FDA, and which has received more than one in-person visit and the "laying on of hands" by Xi Jinping himself. The vice governor and his associates were eager to convince He and Zhang to establish a clinic and research center in Hainan for innovation in reproductive technology and germline genetic engineering. When asked what they wanted—funding, land, a building for their operations—He and Zhang indicated that what they needed above all was a permissive regulatory environment. The officials assured them that they would have it in Hainan.

On their way to Hainan, Zhang and He had first paid a visit to the People's Liberation Army Navy Hospital in Beijing, where fertility services are provided to party officials and others of similar social rank. The chief physician at the hospital is a fertility doctor named Wei Shang. When I interviewed her at the hospital in Beijing in early January 2020, she had just presided over the birth of the first Chinese three-parent baby (to my knowledge, this experiment has never been made public). This birth was the first in what was supposed to have been several dozen, the products of an experimental program of reproductive technology innovation conducted by the People's Liberation Army and based on a recommendation by the Chinese Academy of Sciences for areas of innovation investment where China could get ahead and potentially achieve dominance over the West. Wei had learned the three-parent embryo technique during an extended visit to the US several years earlier, during which she was a resident physician in several fertility labs at major universities. She also worked for several months at Zhang's New Hope Fertility Center. In the aftermath of the He debacle, however, Beijing put a halt to research like Wei's out of concern for further damage to China's reputation. (The production of three-parent embryos technically entails heritable genetic modification, though not through genome editing.)

Wei was positioned to do this work because her research was housed in a military hospital, which was not under the jurisdiction of the Chinese FDA or subject to other regulations on Chinese academic and medical research. She answered directly to the upper levels of the CCP government.

When Zhang and He visited Wei in October 2018, He told her all about the details of his experiment. After she got over her initial shock, she told

He that he should have partnered with her and located his research at her hospital to take advantage of the favorable regulatory situation there. He and Zhang did not immediately accept her proposal because they were exploring their options. From Beijing they went on to Hainan, and then onward to Bangkok. Bangkok was where He had initially located his project until his powerful CCP friends reassured him that he could safely do the work at home—and should, for the sake of his country's "glory" (He's word) and his own.

These anecdotes illustrate the degree to which He operated in an environment permeated by an imperative of innovation married to an ambition for national scientific and technological dominance. The Chinese expressions of that imperative and ambition are by no means unique to China. If anything, Chinese innovation policy is imitative: its political and scientific agendas take their cues from the Western imperative to "win the future" through scientific and technological dominance. He's inhumane project was driven by that imperative. When the leading voices in international science declared He a deviant, they were denying that He was a member of their community and culture. In this effort the Chinese cooperated.

In December 2019, the Chinese government announced that He had been tried, convicted, and sentenced to three years in prison, the maximum applicable sentence under a Chinese statute that imposed criminal penalties for practicing medicine illegally. The statute required a tortured reading to be made to apply to He's actions. (He had sought legal advice before undertaking his experiment and had been assured that his project would not be a violation Chinese law.)

No written record of the investigation or from the criminal prosecution has ever been made public. The prosecution itself violated the most basic norms of due process, including those afforded under Chinese criminal law. He's cooperation in accepting the conviction was secured under threat. The case was narrated to the West by a state-controlled Chinese media outlet, and the news of He's criminal conviction was dutifully parroted around the world by the international media, complete with quotes from prominent scientists declaring the sentence to be just, or even overly lenient. This choreographed show of justice was tailored to satisfy an outraged international scientific community. The criminal had been punished. Case closed.

For better or for worse, the question of criminality is the sovereign jurisdiction of the Chinese courts. But the question of ethical responsibility should not be so hastily answered with casting of stones. The rogue narrative provides false closure: it obscures a scapegoating that placed the stone-throwing mandarins of science squarely on the side of the right, even as He's transgressions opened the door for science to reassert its sovereign authority to govern the future.

Playing Catch-Up in the Race to the Bottom

In 2015, the leaders that convened the first International Summit on Human Gene Editing recognized that the technical issues surrounding risk and safety of HGE were only part of the picture. They rightly determined that the question of whether to proceed required "broad societal consensus" (NASEM 2015).

Less than three days after the He story broke, the Hong Kong summit organizing committee abandoned that commitment and issued a statement reasserting science's jurisdiction over the future, declaring that it is "time to define a rigorous, responsible translational pathway toward [germline genome editing clinical] trials" (NASEM 2018). The presidents of the NAS, the National Academy of Medicine, and the Chinese Academy of Sciences followed up with an opinion piece in *Science Magazine* responding to He's "wake up call from Hong Kong" (Dzau, McNutt, and Bai 2018). The presidents declared that "the ability to use CRISPR-Cas9 to edit the human genome has outpaced nascent efforts by the scientific and medical communities to confront the complex ethical and governance issues that they raise."

This narrative should now be familiar: a new technology raises "complex ethical and governance issues" yet also "outpaces" responses to them. The "scientific and medical communities" who drive forward development of the technology claim the authority and responsibility to respond, even as actions undertaken and encouraged by members of communities "outpace" their own efforts to address those issues. Science first, ethics second.

As we have seen, this story is a self-fulfilling prophecy. It is the story He told himself—and was told by others—to justify his experiment. Yet He's experiment, driven by others' predictions of inevitability, was then itself taken as proof that HGE is inevitable—and thus a *responsible* pathway must be laid for bringing it forth. This is manufactured inevitability. Just as inevitability supplied He with the warrant for self-authorization, so too by treating a future with HGE as a foregone conclusion, the National Academies authorized themselves to lay the "translational pathway" that would bring that future about—and to declare what can make such a pathway "responsible." Thus, manufactured inevitability is used to assert an imperative of governance: science races ahead irrespective of whether societal consensus is or is not achieved. Therefore, governance must cease the futile effort to sail against the winds of scientific and technological change, shift course, and catch up by "thoughtfully crafting regulations of the technology without stifling it" (Doudna 2019). The inevitable must be embraced because it is unavoidable. But it is unavoidable because it is embraced.

In this vision of governance, catching up means ceasing to ask whether science has given us a technological future we wish to welcome or shun. It means acquiescing to what is (putatively) inevitable and then mitigating its negative effects as best we can. Governance—steering the ship—becomes, in this view, an ethically attenuated effort to regain a modicum of control: reclaiming the helm from a mutinous rogue who threatens the future of science, but nevertheless following the course he charted—in effect, doing responsibly what He did irresponsibly. "[The He] case highlights the urgent need to accelerate efforts to reach international agreement upon more specific criteria and standards that have to be met before human germline editing would be deemed permissible" (Dzau, McNutt, and Bai 2018).

Why should this be? In what other domain of social life do we treat the violation of a prohibition as justification for abandoning that prohibition? We do not accept thievery and murder simply because there are always thieves and murderers in any society. Why should we respond to what David Baltimore, in his rebuke to He at the Second International Summit, described as "a failure of self-regulation by the scientific community" by doubling down on self-regulation by the scientific community (Baltimore 2018)? Surely the judgment of an individual scientist, an advisory committee, or even a whole professional community should not be privileged over a societal commitment to democratic governance simply because scientists feel constrained in their freedom to do what they want. Certainly, it is challenging to achieve a democratic and cosmopolitan bioethics sufficient to govern a technology like HGE. But this means that we must redouble our efforts, not abandon them. Far from casting aside the aspiration to "broad societal consensus," we should see with clear eyes what happens when scientists (and, thus, science) are left to decide for themselves the future of humanity.

One of the most startling features of the He debacle is the sense of urgency to race ahead that it has engendered in the upper echelons of international science. It is one thing for a young, ambitious, and reckless scientist operating at "Shenzhen speed" to put the pedal to the floor (Kirksey 2020).

When the leaders of international science declare an "urgent need to accelerate efforts," hitting the gas in hot pursuit of runaway science that they themselves have branded as "rogue," they are recklessly committing to a race of their own making (Dzau, McNutt, and Bai 2018). Why not call a halt instead?

In early 2019, in the wake of the He debacle, some prominent figures did just that. A group led by Eric Lander, who would later become President Biden's Science Advisor, called for a five-year long international moratorium on creating genome-edited babies (Lander et al. 2019). It is revealing that this apparently uncontroversial proposal itself become a focus of significant controversy. Opponents of the proposal objected that the concept of a moratorium is too limiting: it ties scientists' hands (Begley 2019). The objection came down to a preference for self-imposed scientific limitation, so that the scientists who impose limits at one moment are free to lift them in the next. David Baltimore's objection to the proposal illustrates this: "The important point is to be flexible going forward. That's what's wrong with a moratorium. It's that the idea gets fixed in people's minds that we're making firm statements about what we don't want to do and for how long we don't want to do it. With a science that's moving forward as rapidly as this science is, you want to be able to adapt to new discoveries, new opportunities and new understandings. To make rules is probably not a good idea" (Baltimore in Saey 2019).

What is at stake for the opponents of the moratorium proposal is not merely governance of genome editing, but scientific freedom as such—the sovereign authority of science to self-regulate and self-govern. As Baltimore explains: restricting gene editing "could hold back the science" (Saey 2019). In other words, science *should* race ahead, and ethics *should* lag behind—after all, "there's nothing like actually moving ahead [with research] to teach us what the actual pitfalls are" (Saey 2019). He Jiankui made it abundantly clear that the pitfall of blindly and recklessly moving ahead is that one moves ahead blindly and recklessly. But it is a lesson science refuses to learn because to do so would undermine its sovereign self-determination.

Some years have passed, and another International Summit has come and gone. That summit, held in London in early 2023, gave vanishingly little time to the issue of HGE. It began with assurances from Chinese delegates that that country's regulatory house is in order and there shall be no more He Jiankuis. Having ticked that box, a packed, three-day agenda did all it could to avoid straying into the uncomfortable territory of He's experiment and its aftermath. The meeting concluded with yet another statement from the organizing committee declaring HGE out of bounds "at this time," even as it affirmed "legitimate research" and "basic research" that lays the groundwork for making it a reality. When pressed, the organizing committee refused to clarify what distinguishes legitimate from illegitimate research, in effect reserving that judgment for self-regulating scientists.

Meanwhile, He served his sentence, reentered Chinese society, and wasted little time in trying to rehabilitate his professional identity. The media has remained fascinated with the "Chinese Frankenstein," chasing after any scrap of news of his plans, hungry for the least hint of nefarious intent. Initially quiet, He has progressively ratcheted up his visibility, posting on social media and giving media interviews. Although he initially disavowed any intent to work in HGE or any similarly controversial zone, each new iteration of research plans declared on social media has moved him steadily in this direction. In June 2023, He tweeted out an announcement of a newly proposed project, "human embryo gene editing to protect against Alzheimer's disease" (He 2023)—a project that, assuming it does not (yet) entail initiating pregnancies, falls squarely into the Third International Summit's category of "legitimate" basic research. A short while later, He was hired by the Wuchang University of Technology to be the founding director of an institute for genetic medicine.

It should go without saying that the Chinese authorities who are no doubt keeping a close eye on the most infamous scientist in the world could silence He's digital and media presence at any time. That they have not suggests that the man who was once considered an enormous liability to Chinese science is perhaps coming to be seen as an asset. Indeed, a few weeks before I made final revisions on this chapter, the CCP extended an invitation to He Jiankui to join the Party.

Rudderless Governance

For all the summits, advisory reports, and deliberations that have unfolded over the last eight years, there has been remarkably little direct engagement with what strikes me as the most fundamental question of governance of HGE: *What is this technology for?* There has been a tendency to look for scenarios in which embryo editing is "medically necessary" because it is the only means through which someone could have a healthy, genetically related child. These scenarios are then used as justifications for mapping out the conditions under which HGE would be "responsible." Such cases are vanishingly rare. The Nuffield Council (2018) consulted with multiple genetic counselors, who reported never having encountered such a case in what amounted to hundreds of years of combined clinical experience.

Given how vanishingly rare candidate cases are, searching for them amounts to a search for justification, not actual inquiry into the technology's purpose. The designer's intentions do not determine what a technology is for—that is, the reasons it will be used and the uses it will be put to. To understand what a technology is for, one must inquire into the ways it is likely to be incorporated into people's lives and societies. One must look to the ends it might be made to serve and the senses of desire, obligation, or need that usher it in. Once again, the He case offers insight—and a powerful warning.

He's research subjects were married, infertile, heterosexual couples in each of which the man was HIV positive and on antiretroviral therapy (ChiCTR 2018). In China, people living with HIV/AIDS are subject to profound discrimination and often treated as social outcasts (Kirksey 2020). Critics have called He's intervention "medically unnecessary" because there was no need to make the babies immune to HIV, given that there are other, well-established means to avoid transmission of the virus to children. So, what was it for? There is, perhaps, no one who can more honestly answer the question of what this technology is for than those who subjected themselves and their would-be children to it. The following is drawn from a letter written to the judges presiding over He's criminal case by one of the couples that participated in He's clinical trial. The letter responds to the allegation that He coerced them into participating:

> In selecting this type of experiment, we were never misled. It was a form of
> compromise. The object of the compromise was society, and one could even say
> with the whole world. As an AIDS sufferer and family member, we firmly and
> deeply know that it is possible to use a preventative drug to have a healthy
> child. . . . That drug can cure disease, but it cannot cure prejudice. For everyone
> who is listening, please hear this. At a certain level our participation in the ex
> periment was indeed forced, but we were not coerced by any person in particu
> lar. We were coerced by society.
> (Letter from—, translated from Chinese, 2019)

What He offered this couple was, simply, a genetic fix to a social problem: the Chinese problem of intolerable prejudice faced by "AIDS sufferers." What he has offered the rest of us is a lesson in how this technology is bound to be used: not to address the exceedingly rare "medically necessary" cases so central in discussions among scientific elites but to hold out the promise of escape from conditions of human life that are unbearable because societies make them so.

The question of whether the technology can deliver on that promise is unimportant. If people believe that it can, then the technology will be used—and such uses will determine how societies and their members understand what it is for and incorporate it into their children's lives. The project of "consciously perfecting" human life begins with seeking to protect human beings from being marked as lesser, inferior, or defective by "repairing" them. For He's would-be parents, the defect lay not in their future children's DNA, but in a society that sorts, classifies, and discards those persons who it deems to be less-than. No society is free of such defects. Yet acquiescing to them—editing children to compensate for the deficiencies of the societies into which they will be born—doubles down on those deficiencies. Developing and deploying HGE in the name of some unjustifiable notion of "medical necessity" amounts to setting this technology free to be used as genetic fixes for social problems—"cur[ing] prejudice"—of whatever sort the market will sustain. And the market He imagined is vast. Liberating the technology this way would deny most of the human community the freedom to ask if we even want it.

If there is even a small possibility that developing heritable genome editing could contribute to such a future (and as the He case makes clear, that possibility is far from remote), humanity has a duty to own and contend with the question of how to guide and govern this technology. That is a democratic responsibility, one which must be wrested from a would-be sovereign science. Yet the orientation to governance that have guided debate thus far have taken us far off course. Unless we correct that course, the ways of thinking and speaking this essay has examined will continue to drive the direction of technology—encouraging and justifying irresponsible science; sidelining serious ethical reflection; constraining democratic governance; and inhibiting deliberation that draws upon the wealth of human tradition, experience, and moral imagination.

As we navigate into the uncharted waters of our technological future, we would do well to recall, with restraint and humility, that we are an imperfect species. The aspiration to heal the infirmities of our mortal lives, no matter how virtuous, is endlessly vulnerable to that all-too-human impulse to redeem ourselves through mastery of nature and self. It is but a short step from the desire to repair what is imperfect to the impulse to perfect, and the will to perfectibility is a will to power—to render fragile and vulnerable human life an object of manipulation and control.

Beneath a hunger for prestige and a belief in the virtue and righteousness of his project, He Jiankui was also driven by the thrill of it: the reckless speed, the heroic transgression, and the wielding of an unprecedented and seemingly righteous power over the fragility of human life. So too were his aiders and abettors, members of a guild that wields ever-increasing powers over life. They encouraged He not in the name of healing the sick but in the name of an imagined right—and righteousness—of radical scientific freedom to ordain the lives we shall have and the purposes for which we shall live. Far from an aberration, this ethos is a defining feature of twenty-first century technoscience. That ethos produced He Jiankui. It is the wellspring of those forms of science and technology—from the atom bomb to artificial intelligence—that threatens humanity in the name of human progress.

J. Robert Oppenheimer described this ethos more than half a century ago: "When you see something that is technically sweet, you go ahead and do it, and you argue about what to do about it only after you have had your technical success" (US Atomic Energy Commission 1954). The sweet successes of the nuclear age threatened to extinguish civilization itself. Biology's emerging powers over the stuff of life, though far more subtle, are no less threatening to human integrity. It is at our peril that We the People cede the authority to guide and govern our common future to those who would mistake the sweetness of their own success for progress.

REFERENCES

Baltimore, David. 2015. Human Gene Editing Summit: Context for the Summit. December 1, 2015. https://vimeo.com/showcase/3703972/video/149179797. Accessed December 3, 2023.

Baltimore, David. 2018. "Statement of David Baltimore." Second International Summit on Human Genome Editing, Day 2, session 2: Embryo Editing. December 2, 2015. https://livestream.com/accounts/7036396/events/8464254/videos/184103056/playerwidth=640&height=360&enableInfo=true&defaultDrawer=&autoPlay=false&mute=false.

Baylis, Françoise. 2019. *Altered Inheritance: CRISPR and the Ethics of Human Genome Editing*. Cambridge: Harvard University Press.

Begley, Sharon. 2019. "Leading Scientists Call for Global Moratorium on Creating 'CRISPR Babies.'" *STAT*, March 13, 2019. https://www.statnews.com/2019/03/13/crispr-babies-germline-editing-moratorium/.

Belluck, Pam. 2019. "How to Stop Rogue Gene-Editing of Human Embryos?" *New York Times*, January 23, 2019. https://www.nytimes.com/2019/01/23/health/gene-editing -babies-crispr.html.

Bosley, Katrine S., Michael Botchan, Annelien L. Bredenoord, Dana Carroll, R. Alta Charo, Emmanuelle Charpentier, Ron Cohen, et al. 2015. "CRISPR Germline Engineering: The Community Speaks." *Nat Biotechnol* 33 (5): 478–86.

Chinese Clinical Trial Registry (ChiCTR). 2018. Study registration number ChiCTR 1800019378. http://www.chictr.org.cn/showprojen.aspx?proj=32758. Accessed November 27, 2018. Archived at https://web.archive.org/web/20181126073452/https://www .chictr.org.cn/showprojen.aspx?proj=32758.

Cressey, Daniel, and David Cyranoski. 2015. "Human-Embryo Editing Poses Challenges for Journals." *Nat News* 520: 594.

Daley, George Q. 2018. "Current Research and Clinical Applications: Pathways to Translation." Second International Summit on Human Genome Editing, Hong Kong. https:// www.nationalacademies.org/documents/embed/link/LF2255DA3DD1C41C0A42D3B EF0989ACAECE3053A6A9B/file/DEDBE7EF321D645481186F29669759BA177426AD47 8D?noSaveAs=1.

Doudna, Jennifer. 2019. "CRISPR's Unwanted Anniversary." *Science* 366 (6467): 777.

Dzau, Victor J., Marcia McNutt, and Chunli Bai. 2018. "Wake-Up Call from Hong Kong." *Science* 362 (6420): 1215.

Edgerton, David. 2017. "The Political Economy of Science: Prospects and Retrospects." In *The Routledge Handbook of the Political Economy of Science*, edited by David Tyfield, Rebecca Lave, Samuel Randalls, and Charles Thorpe, 21–31. Oxford: Taylor & Francis.

European Group on Ethics in Science and New Technologies. 2016. "Statement on Gene Editing." *Jahr Wiss Ethik* 21 (1). https://ec.europa.eu/research/ege/pdf/gene_editing _ege_statement.pdf#view=fitandpagemode=none.

Evans, John H. 2002. *Playing God?: Human Genetic Engineering and the Rationalization of Public Bioethical Debate*. Chicago: University of Chicago Press.

Greely, Henry T. 2021. *CRISPR People: The Science and Ethics of Editing Humans*. Cambridge: MIT Press.

He, Jiankui. 2018. Medical Ethics Committee Approval Application: CCR5 Gene Editing. Translated by N. Shadid. http://www.chictr.org.cn/uploads/file/201811/bb9c5996d8f -d476eacb4aeecf5fd2a01.pdf. Accessed November 26, 2018.

He, Jiankui. 2023. Post on X. June 28, 2023. https://twitter.com/Jiankui_He/status /1674226970614452227. Accessed June 28, 2023.

Hurlbut, J. Benjamin. 2017a. *Experiments in Democracy: Human Embryo Research and the Politics of Bioethics*. New York: Columbia University Press.

Hurlbut, J. Benjamin. 2017b. "A Science That Knows No Country: Pandemic Preparedness, Global Risk, Sovereign Science." *Big Data & Society* 4, no. 2: 205395171774241.

Hurlbut, J. Benjamin. 2018. "In CRISPR's World: Genome Editing and the Politics of Global Science." In *The Routledge Handbook of Genomics and Society*, edited by Sahra Gibbon, Barbara Prainsack, Stephen Hilgartner, and Janelle Lamoreaux, 169–78. New York: Routledge.

Hurlbut, J. Benjamin. 2019. "Genome Editing: Ask Whether, Not How." *Nature* 565: 135.

Hurlbut, J. Benjamin. 2021. "Decoding the CRISPR-Baby Stories." *MIT Technology Review* 124 (2): 82–84.

Hurlbut, J. Benjamin, Sheila Jasanoff, and Krishanu Saha. 2018. "The Chinese Gene-Editing Experiment Was an Outrage: The Scientific Community Shares Blame." *Washington Post*, November 29, 2018. https://www.washingtonpost.com/outlook/2018/11/29/chinese-gene-editing-experiment-was-an-outrage-broader-scientific-community-shares-some-blame/.

Hurlbut, J. Benjamin, Sheila Jasanoff, Krishanu Saha, Aziza Ahmed, Anthony Appiah, Elizabeth Bartholet, Françoise Baylis, et al. 2018. "Building Capacity for a Global Genome Editing Observatory: Conceptual Challenges." *Trends Biotechnol* 36 (7): 639–41.

Huxley, Julian. 1963. "The Future of Man: Evolutionary Aspects." In *Man and His Future*, edited by Gordon Ethelbert Ward Wolstenholme, 1–22. Boston: Little Brown.

Isaacson, Walter. 2021. *The Code Breaker: Jennifer Doudna, Gene Editing, and the Future of the Human Race*. New York: Simon & Schuster.

Jasanoff, Sheila. 2019. *Can Science Make Sense of Life?* Hoboken: John Wiley.

Jasanoff, Sheila, and J. Benjamin Hurlbut. 2018. "A Global Observatory for Gene Editing." *Nature* 555: 435.

Jasanoff, Sheila, J. Benjamin Hurlbut, and Krishanu Saha. 2019. "Democratic Governance of Human Germline Genome Editing." *CRISPR J* 2 (5): 266–71.

Kirksey, Eben. 2020. *The Mutant Project*. New York: St. Martin's Press.

Lander, Eric S., François Baylis, Feng Zhang, Emmanuelle Charpentier, Paul Berg, Catherine Bourgain, Bärbel Friedrich, et al. 2019. "Adopt a Moratorium on Heritable Genome Editing." *Nature* 567: 165.

Lanphier, Edward, Fyodor Urnov, Sarah Ehlen Haecker, Michael Werner, and Joanna Smolenski. 2015. "Don't Edit the Human Germ Line." *Nature* 519: 410–11.

Liang, Puping, Yanwen Xu, Xiya Zhang, Chenhui Ding, Rui Huang, Zhen Zhang, Jie Lv, et al. 2015. "CRISPR/Cas9-Mediated Gene Editing in Human Tripronuclear Zygotes." *Protein Cell* 6 (5): 363–72.

Ma, Hong, Nuria Marti-Gutierrez, Sang-Wook Park, Jun Wu, Yeonmi Lee, Keiichiro Suzuki, Amy Koski, et al. 2017. "Correction of a Pathogenic Gene Mutation in Human Embryos." *Nature* 548: 413–19.

Narayanamurti, Venkatesh, and Toluwalogo Odumosu. 2016. *Cycles of Invention and Discovery: Rethinking the Endless Frontier*. Cambridge: Harvard University Press.

National Academies of Sciences, Engineering, and Medicine (NASEM). 2015. Organizing Committee for the International Summit on Human Gene Editing. *On Human Gene Editing: International Summit Statement*. Washington DC: National Academies Press. https://www.nationalacademies.org/news/2015/12/on-human-gene-editing-international-summit-statement.

National Academies of Sciences, Engineering, and Medicine (NASEM). 2017a. *Human Genome Editing: Science, Ethics, and Governance*. Washington, DC: National Academies Press.

National Academies of Sciences, Engineering, and Medicine (NASEM). 2017b. *Public Report Release Event for Human Genome Editing: Science, Ethics, and Governance*. Washington, DC: National Academies Press. https://vimeo.com/205455452.

National Academies of Sciences, Engineering, and Medicine (NASEM). 2018. Organizing Committee of the Second International Summit on Human Genome Editing. *Human Genome Editing II: Statement by the Organizing Committee of the Second International*

Summit on Human Genome Editing. Washington, DC: National Academies Press. http://www.nationalacademies.org/onpinews/newsitem.aspx?RecordID=11282018b.

Nuffield Council on Bioethics. 2018. *Genome Editing and Human Reproduction: Social and Ethical Issues*. London: Nuffield Council. https://www.nuffieldbioethics.org /publications/genome-editing-and-human-reproduction.

Polanyi, Michael. 1962. "The Republic of Science: Its Political and Economic Theory." *Minerva* 1, no. 1: 54–73.

President's Commission for the Study of Ethical Problems in Medicine and Biomedical and Behavioral Research. 1982. *Splicing Life: A Report on the Social and Ethical Issues of Genetic Engineering with Human Beings*. Washington, DC: President's Commission.

Regalado, Antonio. 2018. "EXCLUSIVE: Chinese Scientists Are Creating CRISPR Babies." *MIT Technological Review*, November 25, 2018. https://www.technologyreview.com/s /612458/exclusive-chinese-scientists-are-creating-crispr-babies/.

Saey, Tina Hesman. 2019. "A Nobel Prize Winner Argues Banning CRISPR Babies Won't Work." *Science News*, April 2, 2019. https://www.sciencenews.org/article/nobel-prize -winner-david-baltimore-crispr-babies-ban.

Sinsheimer, Robert L. 1969. "The Prospect for Designed Genetic Change." *Am Sci* 57 (1): 134–42.

US Atomic Energy Commission. 1954. *In the Matter of: J. Robert Oppenheimer*. April 13, 1954. Washington, DC: U.S. Department of Energy. https://www.osti.gov/includes /opennet/ includes/Oppenheimer%20hearings/Vol%20II%20Oppenheimer.pdf.

World Health Organization (WHO). 2019. *Expert Advisory Committee on Developing Global Standards for Governance and Oversight of Human Genome Editing: Report of the Second Meeting*. Geneva: WHO. https://iris.who.int/bitstream/handle/10665/341018/WHO -SCI-RFH-2019.02-eng.pdf?sequence=1.

Contributors

FLORENCE ASHLEY, SJD, is an assistant professor at the University of Alberta. A jurist and bioethicist, she served as the first openly transfeminine law clerk at the Supreme Court of Canada and is the author of *Banning Transgender Conversion Practices: A Legal and Policy Analysis*.

NEAL BAER, MD, is an award-winning showrunner, television writer/producer, physician, and author. He is the codirector of the master's degree program in media, medicine, and health at Harvard Medical School. He recently served as executive producer and showrunner of *Designated Survivor* and as the executive producer of *Baking Impossible,* both on Netflix. Baer was executive producer for *ER* and showrunner for the CBS television series *Under the Dome* and the hit NBC television series *Law & Order: Special Victims Unit*.

R. ALTA CHARO, JD, is the Knowles professor emerita of law and bioethics at the University of Wisconsin and has worked for Congress, USAID, and the FDA. She consults for the government, professional organizations, and companies on ethical and regulatory issues related to emerging therapies, reproductive technologies, and bioengineered animals.

MARCY DARNOVSKY, PHD, speaks and writes widely on the politics of human biotechnology, focusing on its social justice and public interest implications. She has worked as an organizer and advocate in a range of environmental and progressive political movements and taught courses at Sonoma State University and California State University, East Bay.

KEVIN DOXZEN, PHD, received his doctorate from the lab of Jennifer Doudna, PhD, at the University of California, Berkeley. His research includes global access to somatic cell gene therapies and ethical considerations for germline genome editing. Doxzen currently works with the Advanced Research Projects Agency for Health (ARPA-H).

ROSEMARIE GARLAND-THOMSON, PHD, is professor emerita of English and bioethics at Emory University and works in disability culture, bioethics, and health humanities. She is the coeditor of *About Us: Essays from the Disability Series of the* New York Times and author of *Staring: How We Look*.

GIGI KWIK GRONVALL, PHD, is an associate professor in the environmental health and engineering department at the Johns Hopkins Bloomberg School of Public Health. An immunologist by training, she leads work to minimize the technical and social risks of the life sciences while advancing biosecurity and biosafety.

JODI HALPERN MD, PHD, Chancellor's Chair and professor of bioethics at the University of California, Berkeley, is an international leader on empathy and the ethics of innovative technologies. Halpern cofounded the Kavli Center for Ethics, Science, and the Public and received the 2022 Guggenheim Fellowship award in medicine.

KATIE HASSON, PHD, writes, speaks, researches, and teaches about the social and political aspects of human genetic and reproductive technologies as the program director on genetic justice at the Center for Genetics and Society. She was previously an assistant professor of sociology and gender studies at the University of Southern California.

ANDREW C. HEINRICH, JD, serves on the faculty of Columbia University's School of International and Public Affairs. He has worked on regulatory affairs matters in government, academia, and industry. He is also the founder and executive director of Project Rousseau.

JACQUELINE HUMPHRIES, PHD, is a research scientist at a biotechnology company headquartered in Emeryville, California. She is a strain engineer and head of her company's R&D infrastructure program.

J. BENJAMIN HURLBUT, PHD, is an associate professor in the School of Life Sciences at Arizona State University. His work examines relationships between science, politics, and law in governance of biotechnologies that raise fundamental questions of human integrity and dignity. He is a codirector of the Global Observatory on Genome Editing.

ELLEN D. JORGENSEN, PHD, is a molecular biologist whose efforts to democratize biotechnology have been chronicled by *Science, Wired, Make, PBS NewsHour,* the Discovery Channel, and the *New York Times.* Her TED talks include "What You Need to Know about CRISPR."

PETER F. R. MILLS, PHD, has been involved for over two decades in the intersection of emerging science, ethics, and public policy. He has worked for the United Kingdom's Human Fertilisation and Embryology Authority, the Department of Health, and the Nuffield Council on Bioethics. He is currently the director of the PHG Foundation in Cambridge, United Kingdom.

CAROL PADDEN, PHD, is a distinguished professor of communication and the Sanford I. Berman Chair in Language and Human Communication at University of California, San Diego and dean of the School of Social Sciences. Her research explores the structure and evolution of sign languages in deaf communities.

MARCUS SCHULTZ-BERGIN, PHD, is an associate college lecturer in philosophy at Cleveland State University. His past work on animal genetic engineering has included "The Dignity of Diminished Animals" in *Ethical Theory & Moral Practice* and "Is CRISPR an Ethical Game Changer?" in the *Journal of Agricultural & Environmental Ethics*.

ROBERT SPARROW, PHD, is a professor in the philosophy program at Monash University, where he works on ethical issues raised by new technologies. He has published on topics as diverse as the ethics of military robotics, the moral status of AIs, human enhancement, stem cells, preimplantation genetic diagnosis, xenotransplantation, and migration.

SANDRA SUFIAN, PHD, is a professor at the University of Illinois–Chicago's Department of Medical Education and in the Department of Disability and Human Development. She cofounded the Cystic Fibrosis Reproductive and Sexual Health Collaborative (CFReSHC), a partnership with the CF female community to identify research priorities and questions about sexual reproductive health and CF.

KRYSTAL TSOSIE, PHD, is an Indigenous (Diné/Navajo) geneticist-bioethicist and assistant professor in the School of Life Sciences at Arizona State University. Her research employs genetic epidemiology, community engagement, policy, and artificial intelligence/machine learning approaches to expand Indigenous data sovereignty, trust, and community data governance in precision health.

ETHAN WEISS, MD, is a cardiologist in San Francisco and an entrepreneur in residence at Third Rock Ventures. There, he helped develop the concept for Marea Therapeutics, where he now serves as chief scientific officer and founder.

RACHEL M. WEST, PHD, is a presidential management fellow in the Waterborne Disease Prevention Branch at the National Center for Emerging and Zoonotic Infectious Diseases, CDC. Her work has focused on *P. falciparum* malaria transmission.

Index

ableist logic, 107
abortion, 156–57; for Down syndrome, 209; termination of pregnancy, 173
acute lymphoblastic leukemia, 174
ADA. *See* Americans with Disabilities Act
Adams, Rachel, 56
Africa: in COVID-19 pandemic, 62; CRISPR in, 203, 204; malaria in, 30, 33
African Medicines Agency (AMA), 203
AGAP005958, 30
Agar, Nicholas, 159–60
aging, 73
Agricultural Marketing Service (AMS), 197
AI. *See* artificial intelligence
albinism, 11–12, 116–23; amblyopia with, 122; compound heterozygote and, 118; genetic testing for, 117
alcoholism, of Indigenous peoples, 137
Alcorn, Leelah, 145
Alzheimer's disease, 238
AMA. *See* African Medicines Agency
amblyopia, with albinism, 122
American Sign Language (ASL), 103–4, 108, 110
Americans with Disabilities Act (ADA), 157, 211
amniocentesis, for Down syndrome, 209
AMS. *See* Agricultural Marketing Service
anemia, 2. *See also* sickle cell anemia
anencephaly, 7, 50
animals: CRISPR for, 9–10, 27, 92–98, 195; disease resistance in, 97; ethics with, 96–98; sentient diminishment of, 95, 96–97; welfare of, 93–96. *See also specific types*
Anopheles gambiae, 30, 33
antibacterials, 29
antibiotic resistance, 29
Ara h 2, 29
ARRIGE. *See* Association for Responsible Research and Innovation in Genome Editing
artificial insemination, 210
artificial intelligence (AI), 78, 84; humility with, 162

Ashley, Florence, 13
Asilomar conference, 35
ASL. *See* American Sign Language
Association for Molecular Pathology v. Myriad Genetics, Inc., 197
Association for Responsible Research and Innovation in Genome Editing (ARRIGE), 80
Atik, Tahir, 105–6
ATTR amyloidosis, CRISPR for, 121
autosomal recessive nonsyndromic hearing loss, 105, 106
avian (H_5N_1) flu variant, 3, 85–86

bacteria: antibacterials, 29; CRISPR for, 85; immune mechanism of, 26
bacteriophages, 25
Baltimore, David, 47, 221, 236, 237
Baltimore Underground Science Space (BUGSS), 33
barcodes, for tumors, 29
Barnard, Christiaan, 227, 228
Baylis, Françoise, 59, 60, 160
beef industry, 94–95
Beltsville hogs, 93, 94
Ben Ouagrham-Gormley, Sonia, 28
beta thalassemia, 1–2, 11
Biden, Joseph R., 61, 237
"Billie Idol," 115, 116
Bioeconomy Research and Development Act of 2020, 197
biohackers. *See* do-it-yourself Bio
biosecurity of CRISPR, 32–37; public notification on, 36–37; recommendations for, 34–37
biosociality, 126, 128–29
bioweapons, CRISPR for, 3, 6, 34–37
blastocyst stage, 214
blindness: albinism and, 119, 121; Leber congenital amaurosis and, 122; somatic gene editing for, 48
breaking the glass, 18–19, 225–31
broad societal consensus: for CRISPR, 18; for HGE, 46–47, 57, 59, 60, 235–36
broiler chickens, 94

BUGSS. *See* Baltimore Underground Science Space

BWC. *See* Convention on the Prohibition of the Development, Production and Stockpiling of Bacteriological and Toxin Weapons and on their Destruction

bystander editing, 136

California Senate Bill 180, 197–98

cancer: activism for, 61; barcodes for, 29; CRISPR for, 10, 29, 89; PGD for, 213

Capecchi, Mario, 167

capitalism, 76; Indigenous, 136

Carroll, Dana, 224

CAR-T cells, 174

Carter, Jimmy, 223

Cas9 enzymes, 1, 26–27, 28, 121, 124, 133, 162; DIY Bio and, 31. *See also* clustered regularly interspaced short palindromic repeats

cats, transgender people and, 143

cattle: in beef industry, 94–95; in dairy industry, 94, 95; dehorning of, 94, 96

CCR5, 72, 74; HIV and, 225

Center for Genetics and Society, 46, 59

CF. *See* cystic fibrosis

CFF. *See* Cystic Fibrosis Foundation

CFTR. *See* Cystic Fibrosis Transmembrane Conductor Receptor

CFTR modulator therapy, 126–28, 129

Charcot-Marie-Tooth, 1

Charo, R. Alta, 17

Charpentier, Emmanuelle, 26, 27, 48, 83, 195

chickens: broiler, 94; debeaking of, 94; dust-bathing of, 97; headless, 10, 95

chimera organisms, 33

China. *See* He Jiankui

cholesterol, 73, 74

Church, George, 4, 8, 11, 26, 27, 53–54, 87–88

civil society, 61

cloning, 53, 211; of Dolly the sheep, 54

closed captions, 110

clustered regularly interspaced short palindromic repeats (CRISPR): in Africa, 203, 204; for animals, 9–10, 27, 92–98, 195; applications for, 29–30; for ATTR amyloidosis, 121; as biosecurity hazard, 32–34; biosecurity recommendations for, 34–37; for bioweapons, 3, 6, 34–37; broad societal consensus for, 18; for CF, 124–29; challenges of, 25–37; community labs for, 9, 83–90; controls, oversight, and monitoring

of, 6, 16–19; criticism of, 195–96; curriculum on, 87–88; for deafness, 103–12; democratization of, 9–10, 37, 83–90; for disabilities, 107–10, 167–68; for diseases, 1–3, 14; diverse voices on, 5; DIY Bio for, 89; efficiency of, 93; for enhancement, 186–87, 195–96; equity and, 17, 200; era of, 5; ethics of, 4–5, 90, 96–98, 142, 163, 199–200, 223; eugenics and, 158, 159, 162, 163, 167; for eukaryotic cells, 27–28; evolution as gene-editing tool, 26–27; FDA on, 86, 201–2; fear of, 17; funding for, 27; future of, 6, 13–16; gene drives and, 89; gene editing and, 133–38; germline cells and, 1, 3–4, 196; goals and boundaries for, 110–12; HGE and, 177; historical development of, 6; for HIV, 195; human rights and, 72, 78; humility with, 129; for Indigenous peoples, 133–38; innovation on, 17; intellectual property rights with, 197; *The Island of Dr. Moreau* and, 9–10; justice and, 17, 200; for Leber congenital amaurosis, 121–22; licensure for, 17; media hype on, 85–87; misrepresentations of, 7–8; for mosquitos, 84; NIH on, 202; off-target mutations with, 4, 28, 93, 172, 200; pathway of, 8–9; personal perspectives on, 5; plasmids and, 26, 29, 87–88; race to the bottom with, 235–38; regulation of, 195–204; research on, 17, 200; safety of, 14, 17, 199; for SCD, 195; self-experiments in, 86, 89, 198; self-governance for, 35–36, 220; somatic cells and, 16; for stem cells, 195; for suffering, 13–14; training for, 17, 37; for transgender people, 139–46; twisted tales of, 45–62; unusual case of, 11; USDA on, 202, 204; WHO on, 204

cochlear implants, for deafness, 74–75, 76

Cohen, G. A., 76

Coller, Barry S., 104, 111

colonialism, 137

Comfort, Nathaniel, 159–60

community labs, 6; for CRISPR, 9, 83–90; gene drives and, 89–90; for IVF, 88

compound heterozygote, albinism and, 118

connexin 26, 26, 105, 106–7, 109

Consensus Study Report of the International Commission on the Clinical Use of Human Germline Genome Editing, 177

Convention on the Prohibition of the Development, Production and Stockpiling of Bacteriological and Toxin Weapons and on their Destruction (BWC), 36

Council of Europe Convention on Human Rights and Biomedicine (Oviedo Convention), 52, 80, 207–8
COVID-19 pandemic, 47–48; HGE and, 61–62
CRISPR. *See* clustered regularly interspaced short palindromic repeats
CRISPR babies, 7, 45, 60, 72, 224, 232; aftermath of, 46–48
critical disability studies, 107–8
crop yields, 29
CrRNA, 1
culture: CF and, 126; deafness and, 104, 108, 111; gene editing and, 215; of Indigenous peoples, 134–35, 137
cyborgs, 144
cystic fibrosis (CF), 11; biosociality and, 126, 128–29; CFTR modulator therapy for, 126–28, 129; CRISPR for, 124–29; culture and, 126; deafness and, 106, 107; DNA replacement for, 127; eugenics and, 157; gene editing for, 126–27; germline editing for, 128; HGE for, 173; identity with, 12, 125–26, 128; obesity with, 127; organoids and, 124; prime editing for, 124, 125; somatic gene editing for, 125; stem cells for, 124; subjectivity with, 125
Cystic Fibrosis Foundation (CFF), 125
Cystic Fibrosis Transmembrane Conductor Receptor (CFTR), 124

dairy industry, 94, 95
Darnovsky, Marcy, 7–8
DARPA. *See* Defense Advanced Research Products Agency
deafness, 11, 47; cochlear implants for, 74–75, 76; connexin 26 and, 26, 105, 106–7, 109; CRISPR for, 103–12; culture and, 104, 108, 111; disabilities and, 104; germline editing for, 105–6, 111–12; goals for, 110–12; heterozygote advantage for, 105–6; HGE for, 75, 104, 111–12, 177; homozygosity and, 105; human rights and, 72; of Jews, 111; off-target mutations for, 106; sterilization for, 107; stigmatization of, 111
Defense Advanced Research Products Agency (DARPA), 33
dehumanization, of transgender people, 13, 142
de Melo-Martín, Inmaculada, 160
democratization: of CRISPR, 9–10, 37, 83–90; of gene editing, 135
de novo organisms, 32

Department of Health and Human Services, US (HHS), 34
Desmond-Hellmann, Sue, 119
diabetes, 54; COVID-19 pandemic and, 62
"The Dignity of Disabled Lives" (Solomon), 162
Direct Genomics, 231–32
disabilities: ADA for, 157, 211; CRISPR for, 107–10, 167–68; deafness and, 104; eugenics and, 55–56, 76, 153–54, 156–57, 162, 167; genetic testing for, 156; HGE for, 5, 74, 79, 175–76; human rights for, 107–10; rights of, with HGE, 59; sterilization for, 55; stigmatization of, 74. *See also specific types*
DIY Bio. *See* do-it-yourself Bio
DNA methylation, 28
DNA replacement, for CF, 127
Dobbs decision, 156
do-it-yourself Bio (DIY Bio), 6, 25, 30–34; biohackers, 10; for CRISPR, 89; FDA on, 198; Jorgenson on, 9–10; STEM in, 37. *See also* self-experiments
Dolly the sheep, 54
Doudna, Jennifer, 26, 27, 28, 48, 83, 195, 224
Down syndrome, 15; abortion for, 209; acceptance of, 209–10; amniocentesis for, 209; genetic testing for, 156; stigmatization of, 181
Doxzen, Kevin, 8–9
dystopia, 55, 85

E. coli, 85
Edison, Thomas, 227
educational attainment, 54, 55; eugenics and, 76
Edwards, Robert, 18, 227–28
egg (gamete) donation, 210; fears of, 209; IVF and, 213
egg industry, 94
Eliot, T. S., 219
embryo selection: albinism and, 122; eugenics and, 156–57; genetic testing for, 173; with PGD, 49, 186; for sex, PGD for, 211
encephalitis, 164
Endy, Drew, 85
enhancement: CRISPR for, 186–87, 195–96; germline editing for, 213; HGE for, 72–74, 186–87
Environmental Protection Agency, US, 90
equity, CRISPR and, 17, 200
Esvelt, Kevin, 3, 5

ethics: with animals, 96–98; of CRISPR, 4–5, 90, 96–98, 142, 163, 199–200, 223; of eugenics, 159; of governance, 236; of He Jiankui, 178–79, 226–27; of HGE, 73–74, 185–90, 222–23; institutional courses on, 36; with transgender people, 139, 142

EU, on GMOs, 198

eugenics: CRISPR and, 158, 159, 162, 163, 167; defined, 155; disabilities and, 55–56, 76, 153–54, 156–57, 162, 167–68; embryo selection and, 156–57; ethics of, 159; gene editing and, 167–68; HGE and, 9, 14, 55–56, 74–77, 157; humane technologies and, 165–68; humility and, 163–65; ideology of health with, 155–58; liberal, 55, 160; limits of human knowledge and, 161–63; modern existential dilemma of, 153–54; new, 158–61; physics of experience with, 161–63; science of, 154–55; social justice and, 74–77; transgender people and, 141; velvet, 56, 153–68

eukaryotic cells, CRISPR for, 27–28

Expert Advisory Committee on Developing Global Standards for Governance and Oversight of Human Genome Editing, of WHO, 32, 80, 183, 187

extinction, of mosquitos, 2–3

FADS1, 137

FAS. See Foreign Agricultural Service

FBI, 33

FDA, 2, 51; on BIO kits, 31, 198; clinical trials and, 52–53, 72; on CRISPR, 86, 201–2; on mitochondrial replacement, 211

fertility doctors/clinics: HGE in, 51, 53–54; revenues of, 53

First International Summit on Human Gene Editing, 7, 18

FMR1, 28

Foreign Agricultural Service (FAS), 202

Fragile X syndrome, 28

function creep, 182

funding, for CRISPR, 27

gamete (egg) donation, 210; fears of, 209

Garland-Thomson, Rosemarie, 13–16, 124

Gattaca, 55

gays, 19; gene for, 140, 141

gender justice, with HGE, 59

gene drives, 29–30; as biosecurity hazard, 32–33; community labs and, 89–90; CRISPR and, 89

gene editing: for CF, 126–27; CRISPR and, 133–38; culture and, 215; democratization of, 135; eugenics and, 167–68; governance of, 215; for Indigenous peoples, 133–38; metaphors for, 163–64; for transgender people, 143

Genetically Modified Organisms (GMOs): EU on, 198; food labeling for, 84, 197

genetic inequality, with HGE, 54–57

genetic testing: for albinism, 117; for disabilities, 156; for Down syndrome, 156; for embryo selection, 173; fears of, 209. See also preimplantation genetic diagnosis

Genius Bank, 210

genocide, 137

GenRich, 55

Genspace, 9–10

germline cells, CRISPR and, 1, 3–4, 196

germline editing: for CF, 128; for deafness, 105–6, 111–12; for enhancement, 213; human evolution and, 213–14; identity and, 188; for Indigenous peoples, 136–37; misrepresentations of, 7–8; moratorium on, 212; prohibition of, 207–8, 223–24; stigmatization from, 209; for transgender people, 140; unknown risks of, 196. See also specific types

germline genome, 2

Ghebreyesus, Tedros Adhanom, 47

GJB2, 105, 106, 107

Global Observatory for Genome Editing, 59, 80

GMOs. See Genetically Modified Organisms

goats, 27

Goldin-Meadow, Susan, 109

Gould, Stephen Jay, 163–65, 168

governance: defined, 220; ethics of, 236; He Jiankui and, 232–41; of HGE, 79–80, 219–41; humility and, 240; lag narrative for, 229–30; rudderless, 238–41; WHO on, 79–80. See also moratorium; self-governance

Gratitude (Sacks), 164

Gray, Sarah, 50

Gray, Victoria, 195

Greely, Henry T. "Hank," 224

gRNA. See guide RNA

Gronvall, Gigi, 6

growth hormone, 93; in tree frogs, 31

Guidelines for Research Involving Recombinant or Synthetic Nucleic Acid Molecules, of NIH, 37

guide RNA (gRNA), 27, 28

H₅N₁ (avian) flu variant, 3, 85–86
Habermas, Jürgen, 160, 190
Hacking, Ian, 155
half-pipe of doom, 85
Halpern, Jodi, 8–9
Hasson, Katie, 7–8
HBB beta-globin, 137
He Jiankui, 3–4, 32, 46–47, 50, 219; affairs of state, 231–38; Alzheimer's disease and, 239; Asilomar conference and, 35; breaking the glass and, 225–31; civil society and, 61; consequences of, 220–21; ethics of, 178–79, 226–27; fertility clinic of, 53; governance and, 232–41; HIV and, 239; house arrest of, 226; humility with, 225; Hurlbut on, 17–19; imprisonment of, 234; international human rights and, 71; jailing of, 45; lag narrative for, 229–30; prestige and, 241; reentry into scientific community, 48, 238; at Second International Summit on Human Genome Editing, 215, 219, 225, 236
Henrich, Andrew, 16–17
heritable genome editing (HGE), 1, 4, 5; alternatives to, 58; broad societal consensus for, 46–47, 57, 59, 60; for CF, 12, 173; challenging implications of, 181–83; clinical trials for, 52–53, 72, 176; commercial dynamics of, 53–54; COVID-19 pandemic and, 61–62; CRISPR and, 177; current debate on, 48–51; for deafness, 75, 104, 111–12, 177; for disabilities, 5, 74, 79, 175–76; for disease prevention, 73–74; for diseases, 11; for enhancement, 72–74, 186–87; ethics of, 73–74, 185–90, 222–23; eugenics and, 9, 14, 55–56, 74–77, 157; fear of, 17, 207–16; in fertility doctors/clinics, 51, 53–54; function creep with, 182; future of, 6, 185–90; genetic inequality with, 54–57; governance of, 79–80, 219–41; for HIV, 186; human rights and, 71–80; hyperbolic claims for, 60; for Indigenous peoples, 136; inevitability of, 224, 227, 235–36; IVF for, 173, 180, 188; limited approval of, 50–51; linear model of innovation for, 221–23; medical model for, 176–77, 179, 181–83; moratorium on, 15, 47, 48, 72, 237; NASEM on, 157; opening conversation on, 57–60; option regret with, 189; pathway of, 8–9; PGD and, 179, 180, 186; prohibition of, 52, 207–8, 223–24; public empowerment with, 59–60; public engagement limitations with, 57–59; public policy on, 52–57; race to the bottom with,

235–38; reconciling risk of, 187–89; responsible pathway for, 176–79; safety of, 8, 14, 187; self-governance for, 236; social justice with, 8, 59, 74–77; social stratification from, 190; therapeutic fallacy of, 174–76; treatment *versus* enhancement with, 72–74; twisted tales of, 45–62; unmet medical need for, 172–83; unusual case of, 11; velvet eugenics and, 56; women and, 56, 78. *See also* He Jiankui
herpesvirus, 31
heterozygote advantage, for deafness, 105–6
heterozygous familial hypercholesterolemia, 177
HGE. *See* heritable genome editing
HHGE. *See* human heritable germline editing
HHS. *See* Department of Health and Human Services, US
Hildebrand, Michael S., 104
HIV, 3, 72, 74, 178; *CCR5* and, 225; CRISPR for, 195; He Jiankui and, 239; HGE for, 186
homozygosity: deafness and, 105; for Huntington's disease, 49, 213; for SCD, 173–74
hormone therapy, for transgender people, 142–43, 144
horsepox, 3
HRIA. *See* Human Rights Impact Assessment
Huang Junjiu, 226
Human Genome Diversity Project, 135
Human Genome Editing: Science, Ethics, and Governance, by NASEM, 223, 228
human growth hormone, 93
human heritable germline editing (HHGE), 79, 208; safety of, 214
human rights: for disabilities, 107–10; foundation for, 77–80; HGE and, 71–80
Human Rights Impact Assessment (HRIA), 71–72, 78–79
humility: with AI, 162; with CRISPR, 129; eugenics and, 163–65; governance and, 240; with He Jiankui, 225
Humphries, Jacqueline, 11, 104
Huntington's disease, 4, 74; homozygosity for, 49, 213; PGD for, 213
Hurlbut, J. Benjamin, 18–19
Al Hussein, Zeid Ra'ad, 78
Huxley, Julian, 223
Hynes, Richard, 223–24

identity: with CF, 12, 125–26, 128; germline editing and, 188. *See also* biosociality; transgender people

IGI. *See* Innovative Genomics Institute
immune mechanism, of bacteria, 26
Indigenous peoples, 13; alcoholism of, 137;
    capitalism with, 136; collected genomes
    of, 135–36; CRISPR for, 133–38; culture of,
    134–35, 137; gene editing for, 133–38; as
    geneticists, 134–35; germline editing for,
    136–37; HGE for, 136; off-target mutations
    for, 136; subjectivity of, 135
Innovative Genomics Institute (IGI), 227
intellectual property rights, with CRISPR, 197
International Commission on the Clinical Use
    of Human Germline Genome Editing, 47,
    208, 214–15; Consensus Study Report of,
    177
International Summit on Human Gene Edit-
    ing, 50, 57, 221. *See also* Second Interna-
    tional Summit on Human Genome Editing;
    Third International Genome Editing
    Summit
intracytoplasmic sperm injection, 188; IVF
    and, 213
invasive species, 32–33
in vitro fertilization (IVF), 10–11, 49, 223;
    community lab for, 88; fear of, 17, 210; fears
    of, 209; for HGE, 173, 180, 188; for PGD, 75,
    173; polygenic scores for, 54; purposes of,
    212–13; women and, 56
in vitro gametogenesis (IVG), 19
IQ, 53–54
The Island of Dr. Moreau (Wells), 9–10
IVF. *See* in vitro fertilization
IVG. *See* in vitro gametogenesis

Jenner, Edward, 227, 228
Jews, deafness of, 111
Johnson v. Calvert, 210
Jonas, Hans, 190
Jorgensen, Ellen, 9–10
Juengst, Eric T., 73
Jurassic Park, 19
justice: CRISPR and, 17, 200. *See also* social
    justice

Kahane, Guy, 161
Kerry, John, 232
Kevles, Daniel, 159–60
Klotho protein, 73
Knight, Tom, 86

lag narrative, 228–30
Lander, Eric, 237

Leber congenital amaurosis, CRISPR for,
    121–22
lesbians, 19
liberal eugenics, 55, 160
licensure, for CRISPR, 17
linear model of innovation, for HGE, 221–23
Liu Huafung, 232
Lovell-Badge, Robin, 224
Lulu, 45

malaria, 2–3, 29; in Africa, 30, 33; gene drives
    for, 34; SCD and, 105
The Man Who Mistook His Wife for a Hat
    (Sacks), 164
Masiyiwa, Strive, 62
mastitis, 95
McKibben, Bill, 56
medical model, for HGE, 176–77, 179, 181–83
Mello, Craig, 232
metaphors: of breaking the glass, 18–19,
    225–31; for CRISPR pathway, 8–9; for gene
    editing, 163–64; of translational pathway, 58
Meyer, Christian G., 105
mice: CF and, 125; CRISPR for, 27; enhance-
    ment of, 73
microRNA (miRNA), 28
Mills, Peter, 8, 11, 15, 58
miRNA. *See* microRNA
Missing Voices Initiative, 59
Mitalipov, Shoukhrat, 226
mitochondrial replacement (nuclear genome
    transfer), 53, 59, 188–89, 211–12; prohibi-
    tion of, 207
modulator therapy: for CF, 126–28, 129; for
    Charcot-Marie-Tooth, 1
Mojica, Francisco, 83–84
monogenetic diseases, 208, 215–16
moratorium: on germline editing, 212; on
    HGE, 47, 48, 72, 237
mosaicism, 196
mosquitoes: CRISPR for, 84; extinction of,
    2–3; gene drives and, 2–3, 30, 34; malaria
    from, 30, 33, 34
mothers. *See* women

NAMSRS. *See* National Academies of
    Medicine and Sciences and Royal Society
Nana, 45
NASEM. *See* National Academies of Sciences,
    Engineering, and Medicine
National Academies of Medicine and Sci-
    ences and Royal Society (NAMSRS), 56,

58; International Commission on the Clinical Use of Human Germline Genome Editing Consensus Study Report and, 177

National Academies of Sciences, Engineering, and Medicine (NASEM), 32, 47; on CRISPR, 103; on germline editing, 212; on HGE, 157; *Human Genome Editing: Science, Ethics, and Governance by*, 223, 228; on human rights, 77; on treatment *versus* enhancement, 72

National Academy of Medicine, 79; on HGE, 47–48

National Academy of Sciences, 79; on HGE, 47–48

National Bioengineered Food Disclosure Standard, 196–97

National Co-ordinating Centre for Public Engagement (NCCPE), 77

National Institutes of Health (NIH), 27, 36, 89; on CRISPR, 202; Guidelines for Research Involving Recombinant or Synthetic Nucleic Acid Molecules of, 37

National Science Foundation, 36

Nature, 47, 77, 79, 226

Nature Biotechnology, 232

NCCPE. *See* National Co-ordinating Centre for Public Engagement

NCOB. *See* Nuffield Council on Bioethics

neurotoxins, 32

New England Journal of Medicine, 121

new eugenics, 158–61

New Hope Fertility Center, 232, 233

New Scientist, 106

New State Ice Co. v. Liebmann, 198

new world screwworm, 30

New York Times, 162, 164

NIH. *See* National Institutes of Health

non-homologous end joining, 27

nuclear genome transfer. *See* mitochondrial replacement

nuclear weapons/power, 3; activism for, 61; Oppenheimer on, 241

Nuffield Council on Bioethics (NCOB), 56, 58, 74; on HGE, 179, 208–9; on HGE governance, 239–40; on human rights, 77, 79

obesity, 55; with CF, 127

OCA2, 118, 120

Odin Kit, 31

Odin Technologies, 37

off-target mutations, 28, 92–93; with CRISPR, 196; for deafness, 106; for Indigenous peoples, 136

1000 Genomes Project, 135

Open Philanthropy Project, 37

Oppenheimer, J. Robert, 241

option regret, with HGE, 189

Organizing Committee, of Second International Summit on Human Genome Editing, 103

organoids, CF and, 124

Oviedo Convention (Council of Europe Convention on Human Rights and Biomedicine), 52, 80, 207–8; on cloning, 211; on mitochondrial replacement, 212

Padden, Carol, 11, 103–4

Parfit, Derek, 188

Path to the Cure, 125

PCSK9, 73, 74

permissibility pillar, 72

Perry, Tony, 224

pest management, 29–30

PGD. *See* preimplantation genetic diagnosis

Phase 4 Regulation, 201

Phipps, James, 228

pigs, 95; Beltsville hogs, 93, 94; CRISPR for, 27

Pinker, Steven, 161

plants, 27

plasmids, CRISPR and, 26, 29, 87–88

polio, 3

polygenic diseases, 208

polygenic score, 54; in eugenics, 55–56

Popescu, Saskia, 28

population bottlenecks, 137

porcine reproductive and respiratory syndrome (PRRS), 95

poultry industry. *See* chickens

pregnancy, 32

preimplantation genetic diagnosis (PGD), 10; after albinism, 119–20; for cancer, 213; for embryo selection, 49, 186; eugenics and, 75; fear of, 17, 210–11; fears of, 209; HGE and, 179, 180, 186; for Huntington's disease, 213; IVF for, 75, 173; for sex selection, 211; unknown impacts of, 188; women and, 56

prenatal testing (PNT). *See* genetic testing

prime editing, for CF, 124, 125

prolepsis, 175

proteins, 29

PRRS. *See* porcine reproductive and respiratory syndrome

public empowerment, with HGE, 59–60

public policy, on HGE, 52–57

Rabinow, Paul, 126
race to the bottom, 235–38
Rebrikov, Denis, 47
recombinant DNA, 223
reprogenetics, 160
research, on CRISPR, 17, 200
Robert, Jason Scott, 9
Roberts, Dorothy, 156
Roe decision, 156
rogue scientists, 3, 6, 14, 16, 19, 160;
    self-governance of, 231. *See also* He Jiankui
Royal Society, 79; International Commission
    on the Clinical Use of Human Germline
    Genome Editing Consensus Study Report
    and, 177. *See also* National Academies of
    Medicine and Sciences and Royal Society
rudderless governance, 238–41

Sacks, Oliver, 164–65, 168
Safe Genes Project, 33
safety: of CRISPR, 14, 17, 199; of HGE, 8, 14,
    187; of HHGE, 214. *See also* off-target
    mutations
Sandel, Michael, 3, 160, 164
Savulescu, Julian, 106
SCD. *See* sickle cell anemia
schizophrenia, 54
Schultz-Bergin, Marcus, 10
Scientific American, 124
Scientific Cooperation Research Program, of
    USDA, 202
*Screening Framework Guidance for Providers of
    Synthetic Double-Stranded DNA*, 34
Second International Summit on Human
    Genome Editing, 178; He Jiankui at, 215,
    219, 225, 236; Organizing Committee of, 103
self-experiments, in CRISPR, 86, 89, 198
self-governance: for CRISPR, 35–36, 220; for
    HGE, 236; of rogue scientists, 231
Sen, Amartya, 76
sex selection, PGD for, 211
Shabot, Sara Cohen, 144
Shearer, A. Eliot, 104
Short, Ada-Rhodes, 139
sickle cell anemia (SCD), 1–2, 11; CRISPR for,
    195; deafness and, 106; *HBB* beta-globin in,
    137; homozygosity for, 173–74; malaria and,
    105; somatic gene editing for, 48; stem cells
    for, 195
sign language, 75, 108–10; ASL, 103–4, 108,
    110
Silver, Lee, 55

Sinsheimer, Robert, 222–23
smallpox, 32
Smith, Richard J., 104
social justice: eugenics and, 74–77; with HGE,
    8, 59, 74–77
Solomon, Andrew, 162, 165
somatic cells, 1, 11; CRISPR and, 16; gene
    editing of, 46; for transness, 13
somatic gene editing: for blindness, 48; for
    CF, 125; for SCD, 48; for transgender
    people, 140, 143
Southern University of Science and
    Technology (SUSTech), 231–32
sovereign science, 230–31
Sparrow, Robert, 15–16
Specter, Michael, 2
Spheres of Justice (Walzer), 76
SRY, 94
STEM, in DIY Bio, 37
stem cells, 2; into animals, 9; for CF, 124;
    CRISPR for, 195; for SCD, 195
sterilization: for deafness, 107; for disabili-
    ties, 55
stigmatization: of deafness, 111; of disabili-
    ties, 74; of Down syndrome, 181; from
    germline editing, 209; of transgender
    people, 142
stroke, 2
subjectivity: with CF, 125; of Indigenous
    peoples, 135
Sufian, Sandra, 12
suicide, of transgender people, 142, 145
super-Mendelian inheritance, 29
surrogacy, 210; fears of, 209; IVF and, 213
SUSTech. *See* Southern University of Science
    and Technology

TALENs. *See* transcription activator-like
    effector nucleases
taonga (treasured), 134
Tay-Sachs, 4, 79
The Telegraph, 124, 128–29
telomerase reverse transcriptase, 73
termination of pregnancy (ToP), 173
test-tube baby, 18, 228
ThermoFisher Scientific, 37
Third International Genome Editing Summit,
    61, 237–38
35delG, 105
three-parent baby, 233
ToP. *See* termination of pregnancy
Tourette's syndrome, 164

tra, 30
training, for CRISPR, 17, 37
transcription activator-like effector nucleases
    (TALENs), 133
transgender people, 6; conversion of, 139–42;
    CRISPR for, 139–46; dehumanization of, 13,
    142; dreams and hype about, 144–46; ethics
    with, 139, 142; eugenics and, 141; high-tech
    medical transition of, 142–44; hormone
    therapy for, 142–43, 144; stigmatization of,
    142; suicide of, 142, 145
trans gene, 140, 141
translational pathway, 58, 235
Traywick, Aaron, 31
treasured (*taonga*), 134
treatment *versus* enhancement, with HGE,
    72–74
tree frogs, growth hormone in, 31
Tsosie, Krystal, 12
tumors, barcodes for, 29

UNCRPD. *See* United Nations Convention on
    the Rights of Persons with Disabilities
UN High Commissioner of Human Rights, 78
United Nations Convention on the Rights of
    Persons with Disabilities (UNCRPD), 157
United States. *See specific topics*
USDA, 196–97; on CRISPR, 202, 204

vaccines: COVID-19 pandemic and, 62; Jenner
    and, 228
vanguard science, 229
vaso-occlusive crises, 2
vector-borne diseases, 30, 39
velvet eugenics, 56, 153–68
Venter, Craig, 224

Viotti, Manuel, 111
viruses, 28

Wallace-Wells, David, 10
Walzer, Michael, 76
Wei Shang, 233–34
Weiss, Ethan, 11–12
Weiss, Palmer, 11–12, 115, 116–17, 119
Weiss, Ruthie, 11–12, 115–23
Wells, H.G., 9–10
West, Rachel, 6
WHO. *See* World Health Organization
WMD Coordinators, 33
women/mothers: eugenics and, 156–57,
    159–60; HGE and, 56, 78; HRIA and, 78
World Health Organization (WHO), 34; on
    CRISPR, 204; Expert Advisory Committee
    on Developing Global Standards for
    Governance and Oversight of Human
    Genome Editing, 32, 80, 183, 187; on gene
    editing, 215; on germline editing, 214; on
    HGE, 47–48, 51, 209; on HGE governance,
    79–80; on public empowerment, 60;
    self-governance and, 35; translational
    pathway of, 58

xenotransplantation, 2
Xi Jinping, 231, 233

YouTube, 86, 89

Zayner, Jo, 31
Zhang, John, 53, 232–34
Zhang, Sarah, 26, 27
Zhu Qingshi, 232
zinc-finger nucleases, 133

HOPKINS PRESS

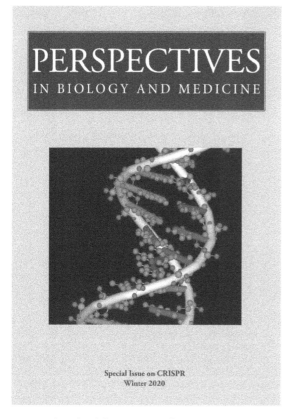

PERSPECTIVES
IN BIOLOGY AND MEDICINE

Special Issue on CRISPR
Winter 2020

Martha Montello,
HARVARD MEDICAL SCHOOL /*Editor*
Olaf Dammann,
TUFTS UNIVERSITY SCHOOL OF
MEDICINE /*Deputy Editor*
Franklin G. Miller,
NATIONAL INSTITUTES OF HEALTH
/*Associate Editor*
Solveig C. Robinson,
PACIFIC LUTHERAN UNIVERSITY
/*Managing Editor*

Perspectives in Biology and Medicine, an interdisciplinary scholarly journal whose readers include biologists, physicians, students, and scholars, publishes essays that place important biological or medical subjects in broader scientific, social, or humanistic contexts. These essays span a wide range of subjects, from biomedical topics such as neurobiology, genetics, and evolution, to topics in ethics, history, philosophy, and medical education and practice. The editors encourage an informal style that has literary merit and that preserves the warmth, excitement, and color of the biological and medical sciences.

Published 4 times per year in Winter, Spring, Summer, and Fall.
P-ISSN: 0031-5982; E-ISSN: 1529-8795.

JOHNS HOPKINS
UNIVERSITY PRESS

www.press.jhu.edu